.TION IN A CHANGING WORLD

'ranklin and Gary McCulloch

'acmillan:

'c High School: Historical Perspectives
and Craig Campbell

leal of Secondary Education

hensive High School?:
, and Comparative Perspectives
.klin and Gary McCulloch

ust Education in American Schools

SECONDARY EDUC

Series editors: Barry M.

Published by Palgrave

The Comprehensive Pu
By Geoffrey Sheringto
(2006)

Cyril Norwood and the
By Gary McCulloch
(2007)

The Death of the Comp
Historical, Contempora
Edited by Barry M. Fr
(2007)

The Emergence of Holo
By Thomas D. Fallace
(2008)

The Emergence of Holocaust Education in American Schools

Thomas D. Fallace

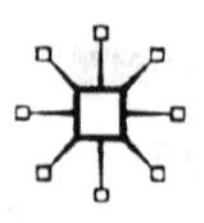

THE EMERGENCE OF HOLOCAUST EDUCATION IN AMERICAN SCHOOLS

First published in 2008 by
PALGRAVE MACMILLAN™
175 Fifth Avenue, New York, N.Y. 10010 and
Houndmills, Basingstoke, Hampshire, England RG21 6XS
Companies and representatives throughout the world.

PALGRAVE MACMILLAN is the global academic imprint of the Palgrave Macmillan division of St. Martin's Press, LLC and of Palgrave Macmillan Ltd. Macmillan® is a registered trademark in the United States, United Kingdom and other countries. Palgrave is a registered trademark in the European Union and other countries.

ISBN-13: 978–0–230–60399–8
ISBN-10: 0–230–60399–8

Library of Congress Cataloging-in-Publication Data is available from the Library of Congress.

A catalogue record for this book is available from the British Library.

Design by Newgen Imaging Systems (P) Ltd., Chennai, India.

First edition: April 2008

10 9 8 7 6 5 4 3 2 1

Printed in the United States of America.

Contents

Series Editors' Preface

Among the educational issues affecting policy makers, public officials, and citizens in modern, democratic, and industrial societies, none has been more contentious than the role of secondary schooling. In establishing the Secondary Education in a Changing World series with Palgrave Macmillan, the intent is to provide a venue for scholars in different national settings to explore critical and controversial issues surrounding secondary education. The series will be a place for the airing and, hopefully, resolution of these controversial issues.

More than a century has elapsed since Emile Durkheim argued the importance of studying secondary education as a unity, rather than in relation to the wide range of subjects and the division of pedagogical labor of which it was composed. Only thus, he insisted, would it be possible to have the ends and aims of secondary education constantly in view. The failure to do so accounted for a great deal of the difficulty with which secondary education was faced. First, it meant that secondary education was "intellectually disorientated" between "a past which is dying and a future which is still undecided," and as a result "lacks the vigor and vitality which it once possessed" (Durkheim, 1938/1987, p. 8). Second, the institutions of secondary education were not understood adequately in relation to their past, which was "the soil which nourished them and gave them their present meaning, and apart from which they cannot be examined without a great deal of impoverishment and distortion" (p. 10). And third, it was difficult for secondary school teachers who were responsible for putting policy reforms into practice, to understand the nature of the problems and issues that prompted them.

In the early years of the twenty-first century, Durkheim's strictures still have resonance. The intellectual disorientation of secondary education is more evident than ever as it is caught up in successive waves of policy changes. The connections between the present and the past have become increasingly hard to trace and untangle. Moreover, the distance between policy makers on the one hand and practitioners on the other has rarely seemed as immense as it is today. The key mission of the current series of books is in the spirit of Durkheim, to address these underlying dilemmas of secondary education and to play a part in resolving them.

In The Emergence of Holocaust Education in American Schools, Thomas Fallace extends our growing body of literature on the Holocaust to the less covered domain of the teaching of this event in American public schools. Fallace's starting point in this volume is the work of American eugenicists, particularly Henry Goddard, during the early years of the twentieth century. It was Goddard and other like-minded Americans, according to Fallace, whose theories of racial hygiene spawned the policies of forced sterilization and immigration restriction in the United States, which inspired the genocidal practices of the Nazi perpetrators of the Holocaust. The American interest in eugenics makes it fitting, according to Fallace, for us to consider how we in this nation addressed their aftermath. His account of the rise of Holocaust education is one part of this story.

In his volume Fallace traces the rise of Holocaust education from a limited effort on the part of Jewish educators during the post–World War II years to raise the consciousness of Jewish Americans about this event to its full inclusion today in the American secondary curriculum. As Fallace sees it, Holocaust education in the United States was a movement launched and sustained by local teachers. He contends in this vein that it was the most successful grassroots, teacher-inspired educational movement in American history. He explores the development of a myriad of curricular programs to teach the Holocaust, the role of television in expanding America's interest in this event, pedagogical developments in the teaching of the Holocaust, particularly in secondary schools, and the appearance on the scene of critiques of this developing educational effort. He concludes his book by looking at the teaching of the Holocaust within the broader domain of history education in American schools and the dangers that the standards movement poses to that aspect of the curriculum.

The Emergence of Holocaust Education in American Schools is the latest effort in this series to consider the important problems affecting contemporary secondary education. Like previous volumes, it employs the lens of historical scholarship to situate pressing issues affecting secondary education within their political and social contexts.

Barry M. Franklin and Gary McCulloch
Series Editors

Preface

The story of the emergence of Holocaust education in American schools takes us from the pages of Jewish education journals to the pages of the *New York Times*, from the New York City Department of Education to the halls of the U.S. Congress, from the classrooms of Brookline, Massachusetts, to the United States Holocaust Memorial Museum. The narrative involves popular Holocaust figures, such as Nobel-Prize-winning author Elie Wiesel, and less-known but equally influential ones, such as Harvard psychologist Lawrence Kohlberg. However, the ambitious educators who launched this movement remain at the center of the story. Although Jewish elites and events in popular culture played a major role in transforming the Holocaust from a peripheral concern to a paradigmatic event, Holocaust education was not simply an inevitable outgrowth of this overall rise of Holocaust consciousness. It was the result of an inspired group of teachers who were responding to a perceived educational need. This is not to suggest that all their ideas were always well-founded; some of their lessons were misguided. Nevertheless, in the era of No Child Left Behind, where politicians and educational policy makers struggle to define the characteristics of a highly qualified teacher, we have much to learn from these innovators, who launched one of the most successful grassroots educational movements in the history of American education. This book is dedicated to them.

This research was the result of two scholarly worlds coming together: historians and educational researchers. These two groups have been working together with more frequency in recent decades as they recognize that an effective history curriculum requires both their perspectives. This trend is encouraging, and I hope it continues. On the history side, I thank Greg Stanton, Jon E. Lendon, and Johann Neem for their encouragement and suggestions. I especially thank Alon Confino, whose enthusiasm for this project since its inception provided me with the confidence I needed to complete it. On the education side, I thank Susan Mintz, Walt Heinecke, and Stephanie van Hover. I also thank Richard Flaim, Rabbi Raymond Zwerin, and Bill Parsons for their cooperation and support during my research. Sam Totten and Simone Schweber were especially helpful by providing me with helpful feedback on many of the chapters. The series editors Barry Franklin and Gary McCulloch also provided me with useful suggestions to improve the manuscript. I thank them both. The support of my father through the years is greatly appreciated. Above all I thank my wife Victoria for her patience and enthusiastic support throughout this entire process.

Much of chapter two appeared previously in the journal *Holocaust and Genocide Studies*. I thank Oxford Press and the United States Holocaust Memorial Museum for permission to use this material. Much of chapter three appeared in *Teachers College Record*. I thank Teachers College, Columbia University, for permission to include the contents of that article.

Introduction

The Story of Vineland, New Jersey

In August 1932 Hanz Kurten of the Nazi Physicians' League organized a course to instruct his doctors in the latest research and theories of racial hygiene. At this meeting and the many others that followed, the Nazi doctors planned their program for racial purification, targeting those whose lives were "not worth living." The program would eventually include the incremental steps of sterilization (eugenics) and the destruction of the mentally ill (euthanasia), before arriving at its sinister conclusion, the state-administered destruction of racial minorities (the final solution). But in 1932, in order to present the most cited and admired research in the field, Kurten did not refer to the work of German doctors. Instead he expressed his admiration for the trailblazing work done in the United States.[1]

Up to that point, the German racial hygiene movement had been largely isolated to the scientific and academic communities. But the Americans were the first to turn theory into practice, using their theories of racial hygiene to inform their sterilization and immigration policies. By the time of Kurten's course, American doctors had sterilized over fifteen thousand "mentally unfit" individuals—many of them while being detained against their will in state institutions.[2] Likewise, the 1924 Immigration Restriction Act had drastically reduced the emigration of Jews and other "undesirable groups" to America.[3] The Nazis considered the Americans pioneers in coping with their genetically unfit and racially inferior population. They would first look to emulate American policies before moving on to more extreme measures.

One of the most influential American eugenicists was Henry Herbert Goddard. The Germans considered Goddard's study of the Kallikak family (published in German translation in 1914) as a classic case and often cited it as evidence of the primacy of heredity over environment. Employing the intelligence test designed by French psychologist Alfred Binet, Goddard identified an intellectual class of feeble-minded people. Goddard suggested that men were born with a general intelligence that was passed along family lines, stating "feeble-mindedness is hereditary and transmitted as surely as any other character."[4] He also thought that morality and emotion were dependent upon this general intelligence. He dubbed those whose intelligence was just below the line of desirability as *morons.*

The morons, according to Goddard, were the biggest threat to humanity because they could potentially escape identification and be allowed to pass on their genes. As he explained, "The idiot is not our greatest problem. He is indeed loathsome ... Nevertheless, he lives his life and is done. He does not continue the race with a line of children like himself... It is the moron type that makes for us our greatest problem."[5] To demonstrate the social repercussions of allowing morons to reproduce, Goddard published a study of some New Jersey paupers he named the Kallikaks. He traced their lineage back several generations to a man named Martin Kallikak, who had married two women; the first a "common tavern wench" and the second an upstanding, "healthy" woman. The ancestral line from his first wife eventually resulted in 480 offspring, the majority of which were morally and intellectually degenerate. The line resulting from his second wife produced 496 offspring who were all fit, healthy, respectable citizens. Thus, according to Goddard, allowing even one moron to reproduce could have terrible consequences. The Nazis looked to Goddard's classic study as proof of how a single "genetic defective" could spoil an entire ancestral line.

Goddard also suggested that morons had certain physical characteristics that could be recognized by the naked eye. In his book he included some crude photographs of the Kallikaks with sunken eyes and darkened mouths. These photographs were later discovered to have been altered by Goddard's pen.[6] Nonetheless, Goddard's study helped to establish the scientifically based notion that one's outward appearance was a reflection of one's intellectual and moral worth. The Nazis shared Goddard's belief that one's moral corruption was manifested in their physical appearance, and concluded that the genetically unfit constituted a major threat to the well-being of the nation.[7] The Nazis looked to prevent the reproduction of those who would, in the words of Goddard, commit the "sin of peopling the world with a race of defective generates."[8]

Goddard never went so far as to recommend the sterilization of these morons. Instead he strongly suggested that they be isolated and prevented from reproducing. The prototype for such an institution was his own research laboratory school, the Training School for Feeble-Minded Girls and Boys in Vineland, New Jersey. While Goddard would have never conceived of the ways in which his research would later be used by the Nazis, the work conducted in this institution (along with that of other American scientists) laid the conceptual groundwork for the eugenic policies of Nazi Germany. The Nazis built upon the genetic research and policy of the United States, fused it with a radical, racist worldview, and implemented the policies that would eventually lead to the final solution. The state-sponsored murder of millions of innocents (mostly Jewish) that took place during World War II under Nazi rule would come to be known as the Holocaust, a word adopted by American and Israeli Jews in the 1960s to describe the event. Many consider the Holocaust the single greatest mass atrocity ever committed against man.

While today it is taken for granted that the average American has a basic understanding of the Holocaust, this was not always the case. Initially in America, knowledge of the Holocaust as an event separate and distinct from the whole of Nazi atrocities was isolated to the Jewish community. By the 1970s awareness of the Holocaust had slowly begun to enter the American consciousness. It wouldn't be until the airing of NBC's miniseries *Holocaust* in April 1978 that millions of Americans would become fully aware of the nature of the Nazi persecution of Jews. The miniseries fueled a growing interest in the Holocaust by non-Jewish Americans that had begun several years earlier. This interest was both initiated and exacerbated by several zealous public school teachers in the Northeast, who began teaching the Holocaust in the mid-1970s as a means of engaging students with the "moral and ethical issues that confront mankind."[9]

Ironically, one of the first schools in the nation to initiate a Holocaust curriculum was in Vineland, New Jersey, the very town in which H.H. Goddard had conducted his eugenicist research half a century earlier. Richard F. Flaim, social studies supervisor from Vineland High School, met with his counterpart Edwin Reynolds of Teaneck, New Jersey, to design a Holocaust curriculum for statewide use. In 1978 their curriculum gained the financial support and endorsement of the Anti-Defamation League (ADL), and in 1982 New Jersey governor Thomas Kean established the Advisory Council on Holocaust Education, which distributed Holocaust curricular materials across the state. In 1994 the state of New Jersey passed a law mandating Holocaust education for every school in the state.[10] By 2007 several U.S. states had imitated New Jersey, passing legislation specifying or mandating Holocaust education, and many more states had developed curricula for voluntary use.[11] Once again the town of Vineland, New Jersey, was the epicenter of a national movement. Through the teaching of the Holocaust, the New Jersey teachers launched a movement to dismantle the very racial stereotypes that H.H. Goddard had helped to establish.

By the 1990s remembering the Holocaust had become a central part of American culture. The Holocaust also emerged as an important topic in the nation's secondary schools. Many states have adopted Holocaust curricula, and there are currently dozens of organizations dedicated to Holocaust study and education in the United States.[12] Most of these organizations offer educational materials, teacher-oriented websites, and access to primary sources. Popular and media interest in the Holocaust came to a crescendo in 1993 with the opening of the United States Holocaust Memorial Museum in Washington, D.C., and the enormous critical and popular success of Steven Spielberg's 1993 film *Schindler's List.* In the wake of this interest historians began to ask questions about how and why the Holocaust had risen to such a high level of prominence in American culture. These studies point to the state-mandated Holocaust curricula as a sign of how the Holocaust has permeated all facets of American culture. Researchers have characterized Holocaust education as a natural consequence of a general rise in consciousness and interest in the event.[13] To a large extent this is true; but Holocaust education in America has its own history, which has not been addressed directly by researchers.[14]

A history of the Holocaust curriculum must address the cultural, popular, and political events that have enabled the topic to ascend to such a high level of prominence in America. Within this context, the present narrative focuses not only on how the topic of the Holocaust emerged in American schools, but also on how the design of Holocaust curricula has changed over time. Curriculum historian Herbert Kliebard writes that "the history of the curriculum focuses on the interaction between curriculum ideas and those political and social conditions that support or undermine their incorporation into the curriculum that is actually experienced at the school level."[15] Social and political contexts must be considered when examining the causes and reception of curriculum, but not to the extent that the historian deprives the curricular designers, such as the New Jersey teachers mentioned earlier, of initiative or agency. Indeed it was the interaction of the curriculum with the larger social, political, and educational forces that have made the Holocaust such a unique and controversial topic, but it was the individual efforts of certain teachers that initiated the teaching of the event.

Holocaust education has been an unprecedented curricular movement for four reasons. First, while the idea of teaching the Holocaust received some initial resistance, it has for the most part been, in the words of sociologist Alan Mintz, "a point of moral consensus between the right and left."[16] Second, while Holocaust education has been a source of *political* consensus, it has been a forum for fierce *curricular* debate. In other words, while everyone may agree that the Holocaust should be taught, they cannot agree on how it should be done. The curricular debate turned into a political one only when state and federal politicians sought to put their stamp of approval on activities that were already taking place at the grassroots level. Third, Holocaust education has endured a major shift in educational and curricular history. Holocaust education in the public schools arose from the neo-progressive strands of the 1960s and 1970s that engendered student "self-realization" and "relevance," but when the political and educational climate changed in the 1980s and 1990s to a more conservative environment, calling for "back-to-basics" and standardized testing, Holocaust education actually gained momentum. In addition Holocaust education, which aligned with the overall goals of the comprehensive high school, increased in popularity at a time when the institution of the secondary school itself came under attack. Fourth, the rise of Holocaust education was a grassroots phenomenon initiated by practicing teachers. While Jewish organizations and state governments often enthusiastically endorsed Holocaust education, the specific curricula themselves were developed by local teachers, who had already been teaching the subject, and often involved the collaboration of Jewish and non-Jewish teachers.

These reasons point to the malleability of the Holocaust as an educational topic. Over the past three decades, political and educational reformers, both liberal and conservative, have molded the Holocaust to accord with their educational and social agendas. Likewise, educators have designed their curricula from a number of different approaches. But all of these approaches have been accompanied by bitter criticism, making the Holocaust one of the most controversial topics in the contemporary American curriculum. Most scholars who have written on the Holocaust in America have ignored this curricular debate, focusing instead on the cultural clash between the Jewish and non-Jewish meanings of the event. This is an oversimplification of the problem. There are many different approaches to the teaching of history, and each of these approaches has been employed to varying degrees in the teaching of the Holocaust. The popularity and moral enormity of teaching the Holocaust has made these different curricular approaches more conspicuous and open to criticism.

The debate over the role of history in the schools predates the emergence of the Holocaust as a topic. Historians, social studies teachers, and educational researchers have been arguing about how history should be taught to American students for over a century. Holocaust education has created a conspicuous forum in which the debate over the teaching of history has been conducted. I argue that the roots of the struggle over the Holocaust in American education do not lie solely in cultural clashes of

America's pluralistic society, but also (if not more so) in the historical roots of the American curriculum. Above all else the Holocaust is history. Therefore, the debate over the Holocaust curriculum should be cast in the context of the role of social studies in the secondary curriculum.

I also suggest that these different approaches to teaching the Holocaust cut across cultural lines. In other words, the debate over whether to approach the Holocaust through a fact-based, affective, or issue-centered progressive manner has been conducted by both Jewish and non-Jewish educators. Examples of each of these curricular approaches can be found *within* each of these groups. But certain Jewish leaders have suggested a particularist approach to teaching the Holocaust, which has been quite problematic for all non-Jewish and certain Jewish educators. The particularists have suggested that the Holocaust is a unique historical event that holds cultural, religious, and metaphysical significance for Jews. They consider a denial of this status by educators as offensive and demeaning. While the particularist approach has not been adopted anywhere outside the Jewish schools, it has been the primary focus of the discourse on Holocaust curricula in America.

Chapter one, "Telling the War," provides a brief overview of the rise of Holocaust consciousness in American culture up to and including the 1960s. This chapter addresses the evolving role the Holocaust played in Jewish education before public schools teacher became interested in teaching the event, and it demonstrates how by the 1970s Jewish educators were divided over how to teach the event to Jewish youth, and how many of these teachers were influenced by the curricular changes taking place in the public schools.

In chapter two, "Holocaust Education in New York City," I explore the circumstances surrounding the first Holocaust curriculum designed for use in public schools, Albert Post's *The Holocaust: Case Study in Genocide* (1973). I then trace the controversy that erupted when the New York City Board of Education attempted to mandate genocide education for all city schools.

In chapter three, "Affective Revolution and Holocaust Education," I describe the origins of an educational curricular movement known as the "affective revolution." I then link this context explicitly to the design of four of the most influential Holocaust curricula in the country: Rabbi Raymond Zwerin and Audrey Friedman Marcus' *Gestapo: A Learning Experience about the Holocaust* (1976), Roselle Chartock and Jack Spencer's *Society on Trial* (1978), Richard Flaim and Edwin Reynolds' *The Holocaust and Genocide: A Search for Conscience* (1983), and William Parsons and Margot Stern Strom's *Facing History and Ourselves* (1982). I argue that the perceived sociocultural crisis of the 1960s engendered teachers' concern for students' individual morals and values, creating a sociocurricular environment conducive for the introduction of the Holocaust into the curriculum.

In chapter four, "Watching and Defining the Holocaust," I discuss how the popularity of NBC's 1978 *Holocaust* miniseries helped launch national interest in the event, and I consider the affect the miniseries had on the emerging Holocaust education movement. However, while the topic of the Holocaust gained popularity in the curriculum and American culture in the late 1970s, a neoconservative climate took hold that threatened the future of genocide education. Against the backdrop of the culture wars, I discuss the controversies that erupted in the 1980s and 1990s over the content of popular Holocaust curricula, including *Facing History and Ourselves* and *Gestapo.*

In chapter five, "Holocaustomania," I discuss why textbooks failed to cover the Holocaust, and how the growing Holocaust historiography supported many of the curricular approaches educators were using to teach the event. I then investigate the origins and design of the Ohio curriculum, *The Holocaust: Prejudice Unleashed.* Based on this curriculum I offer some overall analysis of how and why Holocaust education continued to spread in the 1980s. In chapter six, "Critiquing Holocaust Education," I discuss the United States Holocaust Memorial Museum's "Guidelines for Teaching about the Holocaust" (1993) as well as other published critiques of Holocaust curricula. I explain how William Parsons and Samuel Totten designed the guidelines in response to what they considered the poor quality of existing curricula—including many of the units covered in this very study.

In chapter seven, "Out of the Discourse, Into the Classroom," I explore the rich but recent body of empirical research conducted on Holocaust education. I consider to what degree the Holocaust is being taught, how it is being taught, and what students are leaning from it. In chapter eight, "Teaching the Holocaust and the Aims of Secondary Education," I suggest a way to redirect and assess the conflicting goals of Holocaust education. To do this, I argue, we need to update Lawrence Kohlberg's theory of cognitive-developmentalism—the very theory that helped to launch Holocaust education in the first place. Finally, in the "Epilogue: The Future of Holocaust Education," I suggest that while the Holocaust as a topic is firmly embedded in the American curriculum, the kind of innovative teaching displayed by the educators in my narrative is endangered due to the current movement toward the standardization of the history curriculum.

There has been a tremendous amount of scholarly literature on the meaning and nature of the Holocaust as a historical event and cultural phenomenon. But there has not been a direct investigation of the Holocaust curricula in America and how these curricula were developed. Certain Holocaust curricula, such as *Facing History and Ourselves*, believed by some to be the most popular in America, have been in use for nearly thirty years.[17] One cannot understand how the Holocaust has come to loom so large in American education without exploring its curricular history. While this study will not be exhaustive, it will be an initial step toward understanding how and why teaching the Holocaust has become so popular in America. It will also help to shed some light on the complexities and uses of history in the nation's schools.

Indeed one of the underlying themes of this study will be how for generations professional historians have turned their backs on the history curriculum in the schools. It took an event as controversial and culturally loaded as the Holocaust to ignite their interests once again. And while the opinions of historians were often solicited and encouraged by educational researchers, their suggestions depict how out of touch they are with the reality of teaching history in the schools. Hopefully this study will help to serve as a bridge between the scholarly world of academia and the practical word of the history classroom. At the very least, it will demonstrate how the Holocaust has caused these two worlds to collide.

Chapter 1

Telling the War

On January 30, 1933, Adolf Hitler, leader of the National Socialist German Workers Party (or Nazi party), was named chancellor of Germany. Once in power, Hitler moved quickly to end German democracy by suspending individual freedoms and authorizing special security forces such as the Gestapo, Storm Troopers (SA), and the SS. He also murdered or arrested leaders of opposition political parties. The Nazis, who believed that the racial superiority of the German race was threatened by inferior ones, viewed Jews, Roma (gypsies), and the handicapped as serious biological threats to German racial purity. The German Jews, who numbered about a half a million, were the principal targets of Nazi hatred. The Germans blamed the Jews for the economic depression and for Germany's defeat in World War I (1914–18).

The Nuremberg laws of 1935 made Jews, whom the Nazis defined by the religious affiliation of their grandparents, second-class citizens. Between 1937 and 1939, subsequent German legislation isolated Jews more and more from German society. In November 1938, the Nazis organized a riot known as *Kristallnacht* (the "Night of Broken Glass") against German and Austrian Jews. The attack included the destruction of synagogues, Jewish-owned stores and homes, and the murder of Jewish individuals. During this time, the Nazis targeted other minority groups as well, including Roma, blacks, Jehovah's Witnesses, and homosexuals, and they sterilized over three hundred thousand physically or mentally handicapped individuals.

Between 1933 and 1936, the Nazis imprisoned thousands of political opponents in concentration camps. After *Kristallnacht,* approximately thirty thousand Jewish men were deported to Dachau and other concentration camps and several hundred Jewish women were sent to local jails. In this period, about half the German Jewish population and more than two-thirds of Austrian Jews fled Nazi persecution by emigrating to the United States, Palestine, and elsewhere in Europe. However, most of these countries were unwilling to admit large numbers of refugees.

After the Nazis invaded Poland in 1939, they imprisoned and murdered thousands of Poles to create new living space for Germans; many of them resettled in the area. Following the invasion of the Soviet Union, the Nazis killed tens of thousands of Jews, political leaders, Communists, and Roma in mass shootings by mobile killing squads. The most famous of these sites was Babi Yar, near Kiev, where an estimated thirty-three thousand victims, mostly Jewish, were murdered over two days. The Nazis also killed more than three million Soviet prisoners of war.

During the war, hundreds of new concentration camps were established in occupied territories of eastern and western Europe. In Polish cities such as Warsaw and Lodz, Jews were confined to ghettos where they were subject to starvation, overcrowding, and contagious diseases. Tens of thousands of Jews died in the ghettos. By 1942, the Nazis began deporting ghetto residents to "extermination camps" in Poland to be worked to death and/or immediately gassed. At the infamous Wannsee Conference in January 1941, Nazi leaders decided to implement the "final solution," which meant exterminating all Europeans Jews at six killing sites including Belzec, Treblinka, and Auschwitz-Birkenau. On arrival at these death camps, men were separated from women and children. After being made to undress and hand over all their valuables, the victims were then forced naked into the gas chambers, which were disguised as shower rooms, and gassed by either carbon monoxide or Zyklon B. In the end the Nazis murdered six million European Jews, and millions of other victims.

There were some instances of organized resistance and rescue during the Holocaust. For example, in the fall of 1943, the Danish resistance rescued nearly the entire Jewish community in Denmark by smuggling them to Sweden. Individuals such as Raoul Wallenberg and Oskar Schindler also risked their lives to save Jews and other victims. The most notable example of Jewish resistant took place in the form of an uprising in the Warsaw ghetto in April and May 1943. However, despite knowledge of what was taking place, the American government did not pursue a policy of rescue for victims of Nazism during World War II. It did not increase its quotas for Jewish emigrants and even turned away a ship (the *St. Louis*) of Jewish refugees. American political and military leaders argued that winning the war was the top priority and would bring an end to Nazi terror.

At the end of the war, when Allied troops began to approach the concentration and death camps, the Nazis tried to obscure the evidence of genocide by deporting prisoners to camps inside Germany. Many prisoners died during these "death marches." When the troops arrived, including Allied commander General Eisenhower, the shocked soldiers found piles of corpses and other evidences of what had taken place. After the war, the International Military Tribunal sentenced thirteen Nazi leaders to death in the Nuremburg Trials of 1945 and 1946. In 1948, the U.S. Congress passed the Displaced Persons Act, which provided up to four hundred thousand special visas for Nazi victims. Some sixty-eight thousand of these visas were issued to Jews under the Act, as well as an additional twenty-eight thousand under the Truman Directive during the years 1946–48. Tens of thousands of other Jewish displaced persons emigrated to Palestine.[1]

The Rise of Holocaust Consciousness in America

During World War II, American educators were, on a vague level, aware of the Holocaust while it was occurring. In particular, scholars involved in the intercultural education movement, a reform agenda to help promote pluralism and tolerance in American society, commonly made reference to Hitler's treatment of the Jews. However, these educators were not interested in promoting intervention and rescue missions abroad, but rather they focused their attention on domestic issues of prejudice and discrimination. Intercultural educators were concerned that the war would inspire a backlash against German Americans at home (estimated to be about fifteen million). They also continued to combat American anti-Semitism (Jewish Americans were estimated to be about five million). But above all else, these educators used Hitler's racist policies to draw attention to the plight of African Americans in the South.

Steward Cole, a leading proponent of intercultural education, expressed these tendencies in a 1941 essay. "Take the Jews Lot in Germany" in which he explained, "Adolph Hitler was able to infect with his passionate hatred of the Jews vast numbers of people who were grasping for a recognizable case of their distress." Cole then compared the German hatred of the Jews during the economic depression to the falling cotton prices in the American South, which led poor white Southerners to turn irrationally against their black neighbors. In addition he lamented the rising tide of anti-Italian and anti-German sentiment. He chided citizens who "make no attempt to discriminate between those who are thoroughly American in their loyalties and those who are in sympathy with fascism and National Socialism."[2]

Ruth Benedict, a professor at Columbia University, argued that all American students should learn about discrimination against Jews and blacks at home. "An intercultural program which does not face our Negro problem fairly," she argued in 1942, "would be about on par with a German program which omitted Nazi treatment of the Jews." Domestic prejudice, she implored, "must be faced squarely."[3] In addition, leading intercultural educator Rachel Davis-Dubois authored books titled *The Germans in American Life* (1936) and *The Jews in American Life* (1935); during this period both were considered American minority groups who were being discriminated against to an equal degree. In addition DuBois had worked cooperatively with the American Jewish Committee on a number of projects to reduce anti-Semitism.[4] The Nazi treatment of the Jews in Germany threatened to derail much of the work that intercultural educators had done to develop tolerance of German and Jewish Americans and to assuage tensions between these minority groups. Intercultural educators worked primarily to hold the nation together under the stresses of war.

Immediately after the war, journalists paid a considerable amount of attention to the Holocaust upon its discovery. A poll conducted in May 1945 reported that 84 percent of Americans believed that Germany has slaughtered millions in its concentration camps. In 1944, *Life* magazine published a photographic essay of concentrations camps and the Nuremberg Trials, also covered by mainstream newspapers and magazines, which brought public attention to the numerous Nazi atrocities.[5] While the focus of the trial was not specifically the destruction of the Jews, it introduced the number six million (for Jewish victims) into the public discourse. Douglas Kelly and G.M. Gilbert, who had served as psychiatrists in the trial, even used the term "holocaust" to describe the Jewish experience. "It is up to us," they reflected in their memoirs of the trial, "to determine whether to foster racial hatreds and prejudices . . . whether we learn from the holocaust of Europe and apply what we learn to our own lives."[6] The term holocaust would only be one of many used to describe the Jewish persecution, but it wouldn't become the preferred choice until the late 1960s. At this point, few understood the institutional and ideological enormity of the Jewish persecution. Few understood the centrality of eliminationist anti-Semitism to the Nazi ideology or how greatly the "Jewish question" impacted Nazi foreign and domestic strategies. The number six million, which paled in comparison to the number of overall war-related deaths, seemed to blend into the overall number of Nazi victims.[7]

Following the trauma of the war, Americans returned to prosperity and normality, and quickly shifted their attention from the wartime atrocities of the past to optimistic prospects of the future. Most Holocaust survivors in America, still known as "displaced persons," wanted to get on with their lives and assimilate into American culture. Survivors were often told that few would care to hear about their past anyway. The ideal of emotional triumph in the face of extreme adversity was best embodied in the *Diary of Anne Frank*, first published in 1952. Her diary was later turned into a popular play and motion picture. The first printing of the book sold out quickly and by the end of the year it had sold over one hundred thousand copies. The play was performed more than seven hundred times in cities across the United States.[8]

Both the book and the play *Diary of Anne Frank* were edited for American consumption, which historian Deborah Lipstadt pointed out, "essentially de-Judaized the story by removing many of Anne's own references to her Jewish identity."[9] Her edited diary connected with readers because her account avoided the graphic horrors of the Holocaust and concentrated on the commonality of the human experience. One did not need to be Jewish to identify with her. Her book outlined a general struggle between good and evil, rather than the particularities of the Jewish experience under the Nazis. The entire scope of the Holocaust had not yet been revealed, but scholars were beginning to form a substantial body of research.

During the 1950s there was a steady stream of Holocaust-related scholarship, but it failed to muster much public attention because Holocaust books were often distributed by either Jewish, foreign, or obscure publishers.[10] In 1949 the Conference of Jewish Relations met in New York to discuss the "Problems in the Study of the Jewish Catastrophe."[11] Along these lines, in 1953 Israeli scholars founded Yad Vashem, an institution devoted to the study of the "Disaster," and began issuing bulletins and volumes of research on the Jewish destruction.[12] A few years later, certain researchers started using the term holocaust more consistently, but its use was not widespread. In May 1959, the *New York Times* first printed the term holocaust in reference to the Jewish persecution in its reportage of the dedication of the Yad Vashem memorial.[13]

In the postwar years there was little public interest in Holocaust history. Having been rejected numerous times by publishers, Raul Hilberg published his seminal Holocaust history *The Destruction of the European Jews* in 1961, but only with the subsidy of a survivor family.[14] The exception to the rule was the popularity of William Shirer's *The Rise and Fall of the Third Reich* in 1959, which eventually went on to sell over a million copies. But Shirer's book devoted only a few pages to the destruction of the Jews.

Neither Hilberg not Shirer used the word holocaust to describe the Jewish persecution. By 1960, the term holocaust was only one of many metaphorical/metonymic terms used to describe the Jewish persecution. The words "destruction," "catastrophe," "disaster," or the Hebrew word "shoah" (meaning destruction) were more commonly used.[15] Historians have been unable to trace precisely the point at which the term holocaust became the preferred term in America, though Zev Garber and Bruce Zuckerman attribute its popularity to Elie Wiesel, who unabashedly asserted the religious significance of the event (the Jew as sacrificial offering of mankind) throughout his public life.[16] While this may be overstating Wiesel's influence, it is certain that the term holocaust became more common among intellectuals over the course of the 1960s and that Wiesel was one of many Jewish writers who consistently used this term.

New York City's role as the center of the American Jewish consciousness made it the most obvious choice for launching of Holocaust education on a massive scale. It was also the location where memory of the Holocaust was most conspicuously kept alive. In December 1942, five thousand Jewish workers in New York City stopped work for ten minutes in mourning for and protest over the Nazi slaughter of Jews. Two years later, over thirty thousand Jews memorialized the first anniversary of the Warsaw Ghetto uprising on the steps of New York City Hall, an occasion that featured speeches by Mayor Fiorello LaGuardia and prominent Jewish leaders.[17]

This commemorative tradition continued throughout the 1960s, and in 1970 Mayor Lindsay commemorated the Warsaw battle by designating May 3 "Warsaw Commemoration Day," holding a ceremony in Times Square, and changing its name to "Warsaw Ghetto Square" for the day.[18] In 1972, the Warsaw memorial service was moved to Temple Emanu-El to accommodate the overflowing crowd numbering more than four thousand. The memorial drew support from Governor Rockefeller, who named April 11 as "Warsaw Ghetto Day," and included a message from President Nixon.[19] Although the New York City government enthusiastically endorsed such transient recognitions as commemorative days, it was more reticent to lend support to more permanent memorials. In 1964, a plan for a Warsaw Ghetto Uprising memorial was rejected by the New York City Arts Commission for being aesthetically unpleasing, and for having the potential to inspire other "special groups" to demand their own memorials on public land.[20]

New York was also the headquarters for major Jewish organizations such as the Anti-Defamation League of B'nai B'rith (ADL) and the American Jewish Congress (AJCongress). These groups served as watchdogs for Jewish interests, although the ADL campaigned on behalf of all minority groups, not just Jews. In 1961, the ADL published a critique of the treatment of minorities in textbooks that, along with Jews, included chapters on American Negroes and immigrants. The study did not distinguish the Jewish experience under the Nazis from that of other persecuted minorities. "There is a paucity of forthright material on what Hitler did," the author concluded, "to millions of minority-group members in Germany and the lands he conquered. Seventeen of the 48 textbooks omit all mention of Nazi persecution."[21] The topics receiving the most coverage were Hitler's racist theories, the identity of his victims, including various groups of non-Jews, and the successive stages of brutality, culminating in mass murder.

The topics "most often neglected" included the number of actual victims and the international reaction and consequences of the Nazi reign of terror, such as the Nuremburg war trials. The great majority of textbooks, the author concluded, "do not serve the function that might be expected of them in this area: to present students with a basic overview of the topic of Nazi victimization and slaughter of the vast numbers of innocent people." It is significant to note that the author did not use the term holocaust in his review.[22]

Nine years later the ADL published a follow-up study of minorities in textbooks concluding that "Nazi persecutions of minorities are still inadequately treated."[23] Although the coverage had improved slightly since 1961, a greater percentage of texts were choosing to neglect the topic altogether. "Many textbooks still fail to fulfill their responsibility to American youth," the author explained, and "their failure may be tragic indeed, for those who withhold the lessons of history may be dooming other generations to repeat its mistakes."[24]

By the late 1960s, the rise in ethnic pride made it more acceptable for the ADL to pursue Jewish interests more openly. Distinctions could be made among ethnic hatreds and history. Accordingly, in 1972 the ADL published yet another textbook study by Auschwitz survivor and historian Henry Friedlander. But this time the book was specifically on the Holocaust. He concluded that "What textbook writers are unwilling or unable to do, the interested student will have to do for himself."[25] Through these textbook critiques, one can trace the centrality of the Holocaust to the Jewish consciousness, as it evolved from a subtopic imbedded in a critique of minority representation to a topic worthy of its own annotated bibliography.

The AJCongress also participated in textbook critique, but instead of casting its net to include all minorities, it concentrated on Jewish history. "The presence of Jews in the world from biblical to modern times," one study asserted, "was frequently disregarded, even in such matters as the Hitler atrocities." To combat this deficiency the AJC issued a set of guidelines for book publishers aimed at ending the "distortions and omissions of Jews and Judaism."[26] The AJCongress also furnished a thirty-two-page bibliography to combat what it described as "a lack information about Jewish cultural, spiritual, and historic history in social science textbooks in high schools."[27] While these groups offered critiques, guidelines, and annotated bibliographies aimed at combating textbook deficiencies, they did not engage in curriculum design, nor did they actively pursue an agenda of disseminating materials into the American public schools.[28]

As these studies demonstrate, up to and during the 1960s, the ADL fought on behalf of all minorities as a strategy of protecting Jewish interests. This is not to imply that its work for other American minorities was insincere, but rather that the organization considered anti-Semitism just one form of discrimination. By working to combat a universal prejudice, anti-Semitism could also be contained. While externally American Jews worked to reinforce the authenticity of their Americaness, internally they sought to retain their Jewish culture through education and commemoration. Dealing with the "Catastrophe" was central to this effort.

Jewish Educators Respond to the Catastrophe

Postwar Jewish education in the United States took a number of different forms ranging from full-time Jewish day schools to summer camps, and it incorporated the full range of Jewish movements such as Orthodox, Reform, Conservative, and Reconstructionist Judaism. By the end of the century, the Jewish community had spent close to $1.5 billion on an educational system that included over three thousand schools.[29]

The most significant institutional development in Jewish education was the ascendance of the congregational, or Hebrew, school. These schools became very popular with suburban Jews, especially those in the Reform and Conservative movements. The Hebrew schools provided supplementary education to students attending public schools during the day. They usually met on Sundays and/or on weeknights. Significant to Holocaust education, the popularity of these schools peaked in the 1960s, with approximately 86 percent of Jewish children and adolescents attending. The schools represented a tidy compromise for Jewish parents who wanted their children to assimilate into the populations of their public schools, which were often in elite suburban districts and neighborhoods, but also wanted them to retain and celebrate their Jewish heritage.[30]

Jewish schoolteachers ranged from full-time employees to part-time volunteers. Rabbi Raymond Zwerin described what he called the "paradigmatic" Jewish teacher in the 1960s. She had "a college education, but had been a housewife for 15 years. Her youngest child no longer need[ed] every minute of every day, and so she [had] a little time to teach on Saturday or Sunday."[31] Nonetheless, there were also full-time Jewish schoolteachers who contributed to educational journals and designed curriculum, but often their work was not widely distributed. Overall, American Jewish education was decentralized and diverse. Therefore, it is no surprise that there was a range of ideas on the teaching and meanings of the Holocaust.

Historically, the teaching of the Holocaust in Jewish schools mirrored the role of the Holocaust in Jewish life. In the 1950s, Jewish educators concentrated on the heroic elements of the event as a means of instilling a positive identity in their youth and repairing their fragile Jewish pride. In 1951, Bruno Bettelheim suggested that the primary motivation of Jewish life in America was "to armor the Jewish child against anti-Semitism, to prevent, if possible, the ill effects of his psyche of anti-Semitic experience."[32] One educator suggested that the purpose of Jewish education was to "develop an attitude of ready acceptance of the facts of one's Jewishness . . . [so that] Jews understand and accept their Jewishness as a positive value."[33] Likewise, another suggested that Jewish history should "convey to the child the achievements of his people and for inculcating within him [*sic*] a pride in being a member of the group."[34] In this context, the Holocaust was referred to indirectly and ambiguously, but never confronted directly as a catastrophic event.

References to the Holocaust—which was commonly referred to as the Catastrophe—in Jewish textbooks focused on examples of heroism and resistance. They emphasized that the Nazis never broke the Jewish spirit. One Jewish textbook reported that Jews "were determined to die with dignity rather than submit tamely to the will of a dictator." Similarly, another wrote that the Nazis may have been victors, but the Jews were heroes. One book related the tale of a little boy who used a grenade as a Hannakah light to blow up a Nazi tavern, an act of revenge for the murder of his parents.[35] These tales were more myth than history. However, the heroic approach was not just about putting a positive spin on the horrors of the event. Little was known about the extent and manner of the Nazi assault on the Jews. Survivors mostly kept silent about what they had witnessed and historians had yet to get their minds around the enormity of the event.

In the 1960s, two events in particular helped escalate the Holocaust from a purely academic concern to a larger Jewish one. The first was the capture and trial of former-Nazi Adolf Eichmann in Israel in 1960–61. The trial included the testimonies of dozens of survivors, and garnered a considerable amount of international press coverage. Hannah Arendt reported on the trial in a controversial series of articles for the *New Yorker*, questioning the legality of the trial itself and drawing attention to the Jews' involvement in their own destruction. For many Jews, the trial marked a shift in emphasis from the Jewish heroism of the Warsaw Ghetto uprising to horror experienced by victims of the concentration camps.

The second major event was the Six-Day War in May of 1967, in which the Israeli army quickly defeated Egyptian forces intent on wiping out the Jews. The short-lived threat of total annihilation and international apathy awoke dormant memories of the past. For several tense days, Jews feared that another Holocaust was imminent. As a result of these two events, almost all American Jews were fully aware of the Holocaust and began using it as a point of reference. These events also helped to reintroduce the Holocaust into the consciousness of non-Jews, but only to a limited degree.

These events also sparked a general frustration among older Jews about the lack of concern and knowledge of the event by younger Jews. Rabbi Zwerin summarized the attitude of the generation who witnessed the Holocaust in the following way, "We all know [about the Holocaust] . . . therefore, lets not worry about it . . . sooner of later they [the next generation] will get it through osmosis." By the early 1960s, Zwerin explained, survivors realized that this "osmosis was not happening."[36] A survey of textbooks conduced by the Jewish Community Relations Council of Philadelphia in 1966 confirmed Zwerin's observation. The report concluded the lack of textbook coverage was reflective of "the recent manifestation of a deplorable situation with which Jewish educators have been familiar, namely: the most challenging experience of our time has not been incorporated into the educational heritage of our people."[37]

In addition, during the late 1950s and early 1960s, the historical uniqueness of the Holocaust was beginning to emerge as the historiography matured. Jewish scholars first began comparing the Nazi persecution of Jews to previous pogroms and Jewish persecutions. After this, the next logical step was to compare it to other historic mass atrocities. Raul Hilberg's research outlined the bureaucratic and institutional complexity of the Jewish persecution, which also hinted to its unprecedented nature. The historical uniqueness of the Holocaust was not immediately obvious to scholars, but this perception grew as they began to fathom its enormity.

Lessons of the Catastrophe

When considering American Jewish educators, it is important to understand that many teachers identified themselves as both American and

Jewish and would have prioritized their identities in that order. Likewise, many Jewish educators also considered themselves teachers first and Jews second. Their religious faith did not supercede their ideas about effective teaching, nor did it impact their ideas about how to deliver the content in the most meaningful way. In one of the earliest articles on teaching the Holocaust, Meir Ben-Horin explained this point, "Jewish education is Jewish *education*...it shares with non-Jewish education anywhere in the world the obligations of education as a unique, specific, society-directed process...of deliberate intervention in human growth and human conduct."[38] Therefore, Jewish teachers were influenced, indirectly or directly, by the educational reforms taking place in the public schools, including the "affective revolution" (see chapter three).

However, initially there was considerable resistance to Holocaust education. Shraga Arian reported that in the early 1960s she experienced a general aversion among teachers and parents to teaching about death, a discomfort about the alleged passivity of the murdered Jews, a fear of psychological harm to children, and a general belief that the Holocaust was not relevant for today's teenagers who were antihistorical.[39] Likewise, in the early 1960s, Rabbi Irving Greenberg encountered difficulty selling the idea of a Holocaust course to the Dean of Yeshiva University—one of the nation's most prestigious institutions of Jewish scholarship. He was permitted to teach it only after he agreed to rename the course "Totalitarianism and Ideology in the 20th Century."[40] Gerald Kreefetz in a 1961 article about the neglect of Nazism in textbooks speculated that many publishing companies considered the event too recent a phenomenon to be properly judged, or that it was too controversial and might inspire old hatreds. He also suggested that in light of the Cold War, textbook companies felt uncomfortable recalling the recent past of the West Germany, which was an American ally.[41] In 1972, Saul Friedman interviewed college history teachers about why they neglect to cover the Holocaust, and they responded that they do not get that far in twentieth century, they did not have the expertise, the event was too recent, and that the Holocaust was too unique and did not merit special attention.[42] Nevertheless, in the 1960s, several Jewish educators began to address the darker aspects of the event.

A 1961 article by Meir Ben-Horin reflected this turn. He suggested that it was no longer adequate to tell only part of the Holocaust story. Concentrating on the heroic aspects of the event, he argued, was "an effort to apologize, to falsify through unwarranted prettification of the record." Instead students needed to be confronted with facts of the destruction. "Concentrate on the data" he pleaded. Before one could draw conclusions or even pose educated questions, he argued, the complete scope of the destruction needed to be transmitted, "No Jewish education is Jewish education that fails to enlighten its learners on extermination camps and trains and vans, on illegal ships and marches and flights ... when it fails to provide opportunities for frank, penetrating, critical exploration of the issues involved." Ben-Horin expressed how teachers had a responsibility to enlighten their students with the entire truth, no matter how dark the reality may be. Teachers should draw upon the new historiography, published memoirs, and survivor testimonies to pass along the facts, so that "the intellectual conclusions to be reached by learners belong to them alone and the evidence."[43] He stressed scientific objectivity and emotional detachment when approaching the event.

But not all Jewish teachers agreed with this orientation. In 1964, at its thirty-seventh Annual Conference, the National Council for Jewish Education hosted a symposium on the "Shoah and the Jewish School." The symposium's three speakers demonstrated how, from its inception, there were divergent views on how to teach the Holocaust to Jewish students. The nature of this argument mirrored debates that had been going on about the teaching of history since the progressive era. This debate pitted traditionalists, who called for the transmission of cultural heritage, against proponents of the social studies approach, who called for thematic units organized around student interests, concerns, and problems. While the debate continues unresolved to this day, as we shall see in chapter three, in the 1960s and 1970s, there was a resurgence in the social studies approach, which had a particular focus on "relevant" and "value-laden" material.

At the conference Judah Pilch, director of the National Curriculum Research Institute for the American Association for Jewish Education, argued that the time had come for Jewish youth to be confronted with the "entire story of Jewish martyrdom, including the recent tragedy, the shoah." He asserted that studying the Holocaust would strengthen Jewish identity and expressed apprehension about students' lack of knowledge and concern about Jewish history. "If we fail to impress our children with the Jewish struggle for equality," he feared, "the marginality of their lives as Jews will become greater from year to year." Pilch echoed a growing concern among Jewish leaders that their assimilationist stance of the 1950s had worked too well. American Jews did not instinctually identify themselves with their Jewish ancestors or with the "Jewish people," but rather as "Americans of the Jewish faith." Jewish marriages with non-Jews had further eroded the American Jewish identity. Intensive study of the Holocaust, he suggested, may reverse this erosion and "may even be one of the best antidotes against intermarriage."[44] Pilch aimed to use the Holocaust as means of re-acculturating Jewish youth, and so he stressed the Jewish aspects of the event. While such a position was quite intuitive for Jewish educators, they did not all share his aims. Sara Feinstein, another participant in the symposium, suggested that teachers should emphasize the more universal aspects of the event.

Feinstein was more in tune with the non-Jewish educational research of the time. She quoted John Dewey in her address and asserted that education needed to deal with student values. "It is the task of a good educator," she argued, "to administer the grit and pain in a manner that will encourage proper attitudes of adjustment." Her view pointed to the growing concern among educators with affective learning. She outlined how the Holocaust had become a topic of "immediacy and relevance to our time… not only in the Jewish field, but in the general field as well." Feinstein compared the Jewish experience of the past to the growing racial prejudice of the present, particularly the discrimination against "our fellow Negro Americans." She pointed out how a recent neo-Nazi demonstration in New York City was "not directed against Jews at all, but rather against Negroes."[45] Her comparison to African Americans de-emphasized the particular aspects of anti-Semitism and suggested that the Holocaust was the result of a more ecumenical prejudice. Race was the issue in the Holocaust, she asserted, not religion. Her rationale for teaching the Holocaust was not to acculturate Jews (as Pilch had argued), or to confront them with their past (as Ben-Horin had argued), but rather to speak to the present social conditions—the Dewey-derived idea that was at the very center of the social studies. This was the strand of Holocaust education that would be picked up by non-Jewish educators a decade later. It would also be the approach that traditional Jewish theorists such as Rabbi Zalman F. Ury would reject.

Ury was the third speaker in the symposium, and he expressed a particularist view on the meaning of the Holocaust and why it should be taught to children. He thought the focus of the Holocaust should be its metaphysical significance in Jewish life. Ury asserted that the Holocaust was not only historically but also epistemologically unique; it was a sacred mystery that could never be understood solely through secular, scientific methods. "To propose that we can offer a full explanation of the Catastrophe," he argued, "is presumptiousness equal to one's claim of understanding the essence of G-d." A purely historical explanation of the Holocaust was not appropriate. The event must also be "treated within the framework of our essential theological concepts…to grasp some of the awesome and mystical implications." He felt that the ultimate lesson of the Holocaust was that "the spiritual fiber of the Jewish soul is indestructible." The Holocaust was not the result of a universal strand of prejudice, but rather that of an "anti-Semitism [that] cannot be comprehended."[46] Ury's particularist approach to the Holocaust divorced the event from its temporal and physical context and placed it in a purely Jewish metaphysical one.

Isaac Toubin, executive director of the American Association for Jewish Education refuted Ury's assertion in an article published that year. The Holocaust, he argued, was not "a divine caprice... we should accept it as part of our history and the history of Western civilization... the predictable outcome of man's choice in the eternal struggle between good and evil." Toubin listed a number of lessons to be learned from the Holocaust as a historical event including the universal lesson "of the interdependence and equality of man." Like Feinstein, Toubin found comparisons between the Holocaust and the Civil Rights movement apposite. "In the midst of our American crisis," he asked, "what Jew, understanding the consequences of hatred, can remain indifferent to the plight of the Negro." Educators such as Toubin found that the Holocaust could be made relevant to students' lives by connecting it to contemporary events, which was a central idea of the social studies.[47] In fact, Toubin was not the only Jewish scholar who viewed the Holocaust through a progressive approach.

Connections between the Holocaust and the plight of African Americans appeared in a couple Jewish textbooks, as well as suggestions for affective and experiential pedagogical strategies. The workbook for Ruth Samuel's popular textbook *Pathways through Jewish History* suggested that students partake in a number of role-playing scenarios, seemingly meant to inspire civil disobedience. One situation asked students to imagine how they would respond if they were a member of a German family hiding Jews in their house and a policeman came to the door asking about the missing family. The next scenario, intended to be linked to the prior one, asked students to respond to a similar situation on a segregated bus in the American South. Another scenario asked students to imagine that they were part of an illegal smuggling operation for European Jews to Palestine. The text then asked students how they would respond to a war that they believed was unjust (i.e., Vietnam). Similarly, a teaching guide designed to accompany the popular text *The Jewish Catastrophe in Europe* suggested that teachers make connections between the Holocaust and the Civil Rights movement.[48]

In 1968 Dr. Israel Charny, a clinical psychologist, did not view the events of the Holocaust as a specifically Jewish problem at all. Instead he considered it "one more episode in man's largely unrecognized history as a still primitively-evolved animal who is given to wanton murder." Charny stressed the affective nature of learning about the Holocaust, as "children inevitably learn a good deal about these impulse forces that lurk within them." He specifically attacked the "stick to the facts" traditional approach that failed to address the emotional dissonance the Holocaust created. "To leave sterile facts of our history books as the only thing we teach our children," he argued, "is to fail to encourage an emotionally meaningful or anxiety-responsive experience." He suggested that educators "might try to learn how to invite children to tell their feelings . . . that they find themselves experiencing relative to the subject being studied."[49]

Toubin, Charny, and the workbook authors mentioned earlier demonstrate how the Jewish educational community was not an insular, impregnable group, but rather members of the larger educational community and, therefore, responsive to socio-curricular issues and reforms. Just as non-Jewish social studies teachers began to address the Civil Rights movement, the affective domain, and student values, so did their Jewish counterparts. More importantly, Charny suggested that the Holocaust could be used to challenge students' assumptions and possibly change their future behavior.

Similarly, in 1968, Herman Blumberg called for a Holocaust curriculum that was immediately accessible to student needs, "away from the innocuous and the irrelevant to the burning questions of Jewish youth who seek a proper path through the turbulent last decades of the 20th century." He implored teachers to avoid the "mire of facts" and to curtail "this preoccupation with the past." Instead, he argued, teachers should concentrate on the "genuine, valid questions of Jewish youth." He thought the Holocaust should be used to stir up an emotional response in students, so they can "touch and feel and taste the dark days and burning nights."[50] While the idea of centering the content on student concerns had been a crucial part of the social studies for decades, the preoccupation with students' emotional response was a fairly new educational phenomenon. The social upheaval in American culture in the late 1960s, more so than any Jewish or Israeli events, inspired Jewish teachers to consider their students' values and emotions. Many teachers, both Jewish and non-Jewish, felt a responsibility to address emerging student worries.

On the other hand, certain leaders continued to insist that the metaphysical uniqueness of the Holocaust needed to be explained to Jewish youth. In 1968, Emil Fackenheim attacked the notion of attributing the Holocaust to a universal prejudice. "The most common defense mechanism of the majority of modern Jews," he complained, "is equating the Holocaust with the problem of evil in general. But the Holocaust was a unique event . . . the death of those Jews was unique."[51] Similarly, Alan Bennett asserted "the Holocaust is by its very nature, indescribable."[52] Norman Broznick took this view a step further in 1974, explaining the mystical relationship between the forces of God, the Jewish people, and the Holocaust. He suggested that European Jews might have brought the destruction upon themselves. God chose to punish them collectively because of their eroding Jewish faith. "The estrangement from God on the part of various European Jewries," he wrote "which continued to grow deeper with the passage of time, is the root cause of the calamitous events." He argued that the Holocaust needed to be "integrated into the teaching of prophets, with the holocaust [*sic*] used as a parallel to certain prophesies of doom."[53]

Broznicks's position prompted the response of Marvin Spiegelman, who objected to his suggestion that the Holocaust was a form of divine retribution. He considered this assertion "blasphemous...an aberration and a horror." He then shared an anecdote about a Jewish teenager who was taught that the Holocaust occurred because married women in Europe did not cover their hair, as mandated by Jewish law. He found such teachings irresponsible and potentially harmful to Jewish self-esteem.[54] This particularist view of the Holocaust was not representative of mainstream Jews, but because of the high profile of particularists such as Elie Wiesel, their approach often received a disproportionate amount of attention.

In a more moderate view, Diane Roskies expressed her objections to the progressive and affective approaches to the event in her review of the literature on Holocaust education:

> Relevancy...is debated unceasingly: Teaching for Jewish identification, for faith in Judaism, for "gut reactions," for civil action now, against intermarriage. Quite often the goal of teaching is the *message* of the Destruction rather than the *facts* of the Destruction. In other words the Holocaust is used for Jewish consciousness-raising...Surely the goal of Holocaust education in the Jewish school is not to train children who are intellectual experts on Fascism but children who will be moved by the plight of the Jewish people. But to expect identification as the result of psychodrama is asking too much. (Italics in the original)[55]

Roskies critique contained nothing that pertained specifically to the teaching of the Holocaust as a subject. In other words, her discussion on the goals of history education (message versus facts) could have been about any topic. This debate was simply a continuation of an ongoing argument about the role of history in secondary schools.

Roskies sided with the curricular conservatives and expressed her endorsement of the traditional curricular approach to the Holocaust. "The primary goal of teaching," she explained "should be a full, extensive and complete knowledge of the facts." The facts should take precedence over "future identification with Judaism, Israel, personal security, or ecumenicism between Jews and Christians."[56] Like Ben-Horin, Roskies felt that the focus of Holocaust education should be based on transmitting a thorough description of the event. Other Jewish educators questioned the benefits of bombarding Jewish youth with a depressing narrative. They continued to emphasize some of the positive aspects.

In 1974, the ADL published a textbook on the Holocaust with an introduction by historian Yehuda Bauer, who suggested that all Jewish students "must realize that they, too, are Holocaust survivors." The book was written through the 1950s paradigm of boosting Jewish pride and was entitled *The Holocaust: A History of Courage and Resistance*. Along with the story of the Destruction, the book contained multiple chapters on resistance, rescue, and Jewish warriors. Bea Stadtler, the textbook's author, chose to frame the Holocaust as a persistent triumph, rather than a devastating defeat. On the other hand, other Jewish educators, following the lead of Jewish theologians, continued to accentuate the mystical aspects of the event and framed the event in a specifically Jewish way.[57] The debate over whether to frame the event as secular or mystical continued to engage Jewish scholars in the decades that followed. Throughout the 1960s, while this debate was going on, it never occurred to Jewish leaders that non-Jewish American teachers would ever become interested in teaching the Holocaust to non-Jewish secondary school students.

Holocaust Education in the Public Schools

Despite the increased interest in the Holocaust by Jews, at the beginning of the 1960s mainstream Americans had little to no knowledge of the event. This is not to suggest that they did not know that Jews were persecuted and murdered by the Nazis. In fact, there were numerous references to the Holocaust in television shows, films, and literature in the 1950s and 1960s that depended upon this basic knowledge.[58] But Americans did not know the extent, scope, and nature of the persecution. They did not think of it as an event distinct and separate from the overall Nazi assault on Europe. In other words, they did not know all the aspects of the Holocaust that distinguished it from all other atrocities. Even young assimilated Jewish Americans had only limited knowledge of the event, and this was learned largely from television. "School didn't explain the murder of the Jews," Steve Cohen, a Massachusetts teacher who grew up in Queens, recalled, "and mainstream culture didn't usually mention it... [but] Mel Brooks and Woody Allen referred to it all the time in funny ways... Nazis showed up in all kinds of movies, war games and even sitcoms." Despite growing up in an American Jewish community, Cohen had not even heard the word Holocaust in reference to the Jewish persecution until he graduated from college in 1972.[59] Over the course of the 1970s, Holocaust survivor and award-winning author Elie Wiesel would help to raise the profile of the event substantially.

For most Americans Elie Wiesel is the symbol of Holocaust memory. He has written and spoken prolifically on the Holocaust for over forty years. Wiesel's rise in popularity and status mirrored that of all Holocaust survivors, and he often spoke on their behalf. But this was not always the case. For a decade after his own liberation, Wiesel refused to discuss his past in the Nazi death camp until a particular incident inspired him to break his silence. After the war, Wiesel began working as a correspondent for an Israeli newspaper, leading to an interview with Nobel-Prize-winning author Francois Mauriac. During this interview, he entered into a heated discussion with the author about Christian love in which Wiesel, in refutation, shared his experiences in the concentration camp. Mauriac encouraged Wiesel to speak out about the horrific things he witnessed. In 1956, Wiesel followed this advice and published his 245-page Holocaust memoir, *And the World Remained Silent,* in Yiddish. Two years later, he published a 178-page French version entitled *La Nuit.* In 1960, after having been rejected by numerous publishers, an American version of *Night* was finally published with an introduction by Mauriac. It barely sold a thousand copies. In time *Night* would become the most widely read Holocaust memoir in the world.[60]

In November of 1972, the *New York Times* published an article by Wiesel entitled "Telling the War." Wiesel, then a professor at City College of New York, reviewed several Holocaust-related books for children. He expressed how for years Holocaust survivors "did their best to shield their children from a subject they considered too depressing."[61] Holocaust survivors had left the past behind and moved on with their lives. They didn't think anyone would listen or understand. "Suddenly," Wiesel wrote, "the situation has changed. The theme of the Holocaust was no longer taboo. It is now discussed freely." He suggested that this sudden interest was fueled by the fact that Jewish children were now separated from the event by an entire generation, and that there was a growing overall interest in Jewish history. Wiesel was encouraged by these developments and hoped that more survivors would "speak up, to translate what they have endured-to speak to children of things that disgraced adults." He pleaded that all children, not just Jewish, "be exposed to yesterday's grief and memories which, unbeknownst to them, are part and parcel of their daily experience."

Wiesel's article, more than any other document, marked the beginning of the movement toward Holocaust education in American public schools.

Holocaust educators would cite this article as a rationale for introducing the topic to their students. "Young people today don't want to be shielded," Wiesel suggested, "They want to learn about the heinous kingdom where, long ago, the young were not allowed to live-and neither were the old."[62] The first teachers of the Holocaust agreed.

In the early 1970s, there was also a rising interest in the Holocaust among colleges and universities. Franklin Littell initiated the first American graduate seminar addressing the Holocaust at Emory University during the 1959–60 school year. Littell, a Methodist minister, had served as an educational officer in the American occupation of Germany. The next year Marie Syrkin at Brandeis University and Yaffa Eliach at Brooklyn College offered the first American undergraduate courses on the Holocaust.[63] In 1972, the *New York Times* reported that a new student-designed course on the Holocaust would be offered at Hampshire College "largely in response to student urgings for reform and 'relevance.'"[64] They wanted such a course because, according to one student, "the destruction of six million innocent people was a uniquely tragic event in Jewish and human history, but nothing about it was ever taught to us in our public schools." The students resented the older generation, who had denied them access to knowledge about the Nazi atrocities. One student claimed that "the present generation of students is notably a post-Holocaust one for whom that event is remote and unreal... knowledge and study of the event has been confined to the scholars."[65] Likewise, Holocaust courses had been positively received at colleges such as Colgate and the University of Bridgeport. A university spokesman for Bridgeport said, "the response has been tremendous, so large that we have had to turn some people away."[66] In 1975, Franklin Littell, then chairman of religious studies at Temple University, estimated that over a hundred institutions were teaching some kind of course or seminar on the Holocaust.[67]

However, by the early 1970s, interest in the Holocaust was still largely isolated to the scholarly and Jewish communities. The historiography was still relatively small and unread. In 1972, in a review of Holocaust literature, Gerd Norman concluded "that there is no Holocaust phenomenon in the historical writing... in the United States, except among practitioners of Jewish history and Jewish intellectuals."[68] Non-Jewish Americans had little to no scholarly interest in the Holocaust. Likewise, in the early 1970s, the term holocaust would not have conjured up the destruction of the European Jews in the minds of most Americans. For example, in March of 1970, rock critic Nik Cohn wrote a review of the new album by the energetic British rock group The Who. "*Live at Leeds*," he wrote, was "the best live rock album ever made... the definitive hard rock holocaust."[69] It wouldn't be until the 1980s that the word holocaust would be indelibly linked to the destruction of the European Jews in a way that would have made Cohn's casual use of the word inappropriate.

The Holocaust became a topic of interest for secondary school teachers in the mid-1970s. Several teachers simultaneously, but independently, arrived at the conclusion that the neglect of this important topic had gone on long enough. Drawing on contemporaneous examples of genocide in the news and upon emerging educational theories, these teachers introduced the Holocaust as a way to engage students in a topic of relevance and to encourage them make immediate connections to their lives. The first Holocaust curriculum intended for distribution in public schools was published in New York City in 1973. The designer directly cited Eile Wiesel's *New York Times* article as a rationale for his unit. The design and reception of this curriculum will be the subject of the next chapter. However, a far more influential group of curricula were designed and implemented in Massachusetts and New Jersey. I will address these in chapter three. Taken together these curricula, each designed at the grassroots level, launched the Holocaust education movement, which continues to grow to this day.

Chapter 2

Holocaust Education in New York City

When the Holocaust was first introduced into the American public secondary schools, knowledge of the event was mostly limited to areas with large Jewish populations. Therefore, it is no surprise that New York City was the first location where the widespread Holocaust education began. It is difficult to say with any certainty who the first teachers of the Holocaust in American public schools were. There seemed to be several social studies teachers who began teaching a Holocaust unit around 1972–73. Many teachers most likely taught Holocaust literature before that (e.g., Leatrice Rabinsky; see chapter five). In addition, there were likely many Jewish public school teachers throughout the country who would have had more than a superficial knowledge of the event prior to 1973, who may have taught the Holocaust in some manner. Teaching the *Diary of Anne Frank* would have been the most common example. For this reason, the reader must pay attention to textual distinctions, such as "course," "unit," and "mandatory." The impetus for the first Holocaust units designed for use in public schools can be traced to a large degree to the work of Holocaust survivor and author Elie Wiesel.

Wiesel moved to New York City in the late 1950s. Despite the initially lukewarm reception of *Night*, his writing was beginning to receive more attention by critics and the general public. In 1966, he received a literary award from the ADL and his work was being discussed in Jewish journals.[1] In November 1975, Wiesel wrote another article devoted to teaching the Holocaust, which appeared in *New York Times*. In it, he shared his own experiences teaching the event to his college students, many of whom were the children of Holocaust survivors. "They know, as I do," he wrote, "this is a course like any other course. They see it [as] a kind of initiation. Almost all emerge transformed-more human, and more anguished too." He reported how his students were "stormy and passionate" and how the Holocaust pressed "their identities to the limit." While his colleagues complained that students had recently become less disciplined and apathetic, Wiesel found the exact opposite. "Intelligent, alert," he exclaimed, "my students never cease to astonish me with their thirst to know."[2] Wiesel asserted that the Holocaust could be a transformative topic for students.

Wiesel's Holocaust class answered the students' call for relevance, because it not only transmitted the facts of the event, but, as Wiesel pointed out, it also spoke to their "identity" as American Jews. The Holocaust was relevant to these students because they were Jewish; it gave them a sense of common history. In the cultural context of the 1970s, the celebration of cultural difference was appropriate and encouraged. Black nationalists had been attacking the "melting pot" paradigm for years; they inspired other ethnic groups to look inward and discover what distinguished them from their peers. At this point, it never occurred to Jewish educators that the Holocaust would appeal to non-Jewish students as well. But the relevance of the Holocaust for non-Jews did not lie in its Jewish particularities. Instead the relevance lay in its connection to the term genocide—a general term used to describe the systematic, planned annihilation of a racial, political, or cultural group.

The Relevance of Genocide

The term genocide had reentered the public discourse in the late 1960s in reference to a number of group atrocities, which literally brought the term to the front page of many newspapers. Since, according to the 1948 Convention of the Prevention and Punishment of the Crime of Genocide (Genocide Convention), the United Nations (UN) was legally bound to intervene in genocides, political leaders from around the world used the term to draw attention and support for their particular situation. In 1969, El Salvador accused Honduras of genocide against the Salvadorians living within their borders, and in Northern Ireland, three members of parliament accused the Ulster government of genocide against Irish Catholics. In 1970, the Palestine Liberation Organization charged King Hussein of Jordan with genocide, Biafrans accused Nigeria of genocide, and Indian Muslims accused Hindus of genocide in India and Bangladesh. In 1972, the Sudanese liberation movement charged the Arab-led government of

Sudan with the crime. The term genocide, originally conceived to describe the Holocaust, naturally made reference to the atrocities conducted by the Nazis. Leaders evoking the term suggested that if international action was not taken, a similar catastrophe could happen again. Leaders often used the term loosely but freely because the word could potentially cut through the malaise of international apathy.[3]

Domestically the term was gaining attention as well. A 1969 book by Jean-Paul Sarte, *On Genocide,* accused the U.S. government of genocide in Vietnam. That same year the North Vietnamese made a similar accusation. In 1970, the Black Panther Party threatened to charge the U.S. government with genocide against American blacks, while the newly formed Black Rights group "Making the Nation" organized a march from Maryland to New York to bring genocide charges before the UN. But the most conspicuous domestic event surrounding genocide involved an attempt by President Nixon to encourage the U.S. Senate to ratify the 1948 Genocide Convention. While seventy-five nations had ratified the agreement, the United States had not, stating that such international laws could potentially interfere with federal and state jurisdiction. In February of 1970, having been prompted by Britain's ratification of the treaty the week before, President Nixon urged Congress to reconsider. However, Nixon's movement was squashed after a revote by the American Bar Association (ABA) reaffirmed its 1949 position against the pact. The ABA worried that the United States may be brought up on genocide charges for their policies in Vietnam.[4] Thus, a series of events, both national and international, brought the concept of genocide to the attention of the general public, which indirectly brought more attention to the Holocaust itself. How could one talk about genocide without referencing the Holocaust? The numerous examples of potential genocides across the globe proved that the world had not learned its lesson.

This rise in genocide consciousness crossed with another ideological current called the "affective revolution," which will be explored in the next chapter. Drawing upon these ideas, secondary school teachers quickly realized the pedagogical potential of a unit organized around genocide, and how it could engage student interest and connect with student concerns. Of course, any unit of genocide would have to be centered on the destruction of the European Jews—the archetypal example of the concept. Therefore, teachers plucked the numerous examples of contemporary genocide from the front page of newspapers and compared them to the Holocaust, not as a way to "Americanize" the event, but as a way to connect the events of European history to the present. If international leaders compared contemporary atrocities to the Holocaust, teachers asked, why shouldn't they? Designed on the heels of events mentioned earlier, Albert Post, assistant director of social studies for New York Public Schools, designed his unit as a case study in genocide. He asked students to consider whether the Holocaust actually did resemble the events to which it was being compared.

The Holocaust: A Case Study in Genocide

The 1973, the curriculum *The Holocaust: A Case Study Of Genocide* was published by the Commission on Jewish Studies in Public Schools of the American Association for Jewish Education "in response to a wide demand by public schools for supporting curricula on the subject of genocide." Its author, Albert Post, undertook the project "as a community service."[5] Post's curriculum was very practical and malleable in design, showing his direct knowledge of the realities of the classroom. The unit contained only five lessons, designed to be fused into social studies courses in grades seven–twelve dealing with World War II. He also provided suggestions and materials for a "mini-course" for those teachers who chose to spend more time on the topic. For its first reading, Post's curriculum included Elie Wiesel's 1972 *New York Times* article on the teaching the Holocaust as a justification for why students needed to learn about this event. Wiesel's article spoke directly to the tenuous nature of humanity's goodness. He deliberately concentrated on the Jewish experience and did not make any reference to contemporary problems such as the Civil Rights movement or Vietnam. However, these connections would be made by Post.

The curriculum presented the Holocaust as the most extreme case in history of the phenomenon of genocide. Despite this thematic approach, Post centered his curriculum on the historical facts and chronology of the Holocaust. But he did freely make connections with contemporary events. Genocide was a word not only to be applied to the Jews, Post argued, but it "should also bring to mind the premeditated, ruthless official campaign of the Turkish government" as well as "Biafra, where millions were made to starve during a war of succession" and "Bangladesh where millions were made homeless, forced into exile and . . . perished from starvation and mass murder."[6] He didn't seem to think that this approach contradicted Dr. Hyman Chanover of the American Association for Jewish Education, who in the foreword to the curriculum declared the Holocaust "unprecedented in the annals of history."[7] To make a case for Armenian genocide, Post skillfully chose an excerpt from a Jewish author named Stanley Rabinowitz, comparing the similarities between the Holocaust and the

Armenian massacre: "Like the Jews the Armenians are a people who uphold a religious civilization, embracing nationality, culture and faith... Like the Jews, the Armenians are an ancient people... Like the Jews, the Armenians have preserved their identity through a long series of invasions, subjugations and dispersion."[8]

Accordingly, Post's course objectives were organized to progress from the particularities of the Holocaust to the universality of genocide in modern culture. He wanted students to "gain understanding of the unprecedented nature and scope of the deliberately planned annihilation of the Jewish people," and how the Nazis used "Anti-Semitism as their primary propaganda instrument," as well as "realize that genocide is a threat to all humanity" and "inspire a present generation of youth to help build a world in which genocide shall not again occur."[9]

Post's language spoke to the curricular zeitgeist of relevance and affective learning. The "pivotal" question to be posed throughout the unit, Post suggested, was "whether or not the United States was guilty of genocide in North Vietnam because of the massive bombing if the cities." He also suggested that a "mix of affective and cognitive strategies" would have a "more stimulating impact," and that learning about specific genocides would enable students, "regardless of race, creed or ethnic group," to draw out their "universal and contemporary implications." While the five-lesson unit was designed for instructing all students, his suggested three–eight-week mini-course was far more focused on the Jewish experience, including sections on "Jewish Life in Europe in the 19th Century," "Jews in Battle," and it ended with "Israel: The Great Homecoming." This demonstrated his understanding that public school teachers who would be teaching the long course would likely be instructing large Jewish populations, or even teaching the Holocaust as an elective to an all-Jewish class.[10]

Overall, Post's unit was typical of what was to come. It showed the awkwardness of having to balance the particularities of the Holocaust with the universality of discrimination—of teaching the Holocaust as an "unprecedented" event "and as a case study in genocide." Compared to later Holocaust units, Post's curriculum was weak in its conceptual framework because he did not ground his lessons in the research and theories on moral reasoning or the recent work of social studies researchers on value-conflict (explored in chapter three). His only rationale was the 1972 *New York Times* article by Wiesel. He did manage to avoid any sweeping statements about racism, intolerance, or the nature of man. In this sense, his approach was traditional as opposed to progressive because he concentrated on the facts and historical particularities of the specific events. His lessons were meant to be implemented seamlessly into a unit of World War II, which would have stressed its historical context. The discussion on other genocides and Vietnam were meant to be peripheral; the center of the unit was the historical facts of the Jewish experience under the Nazis.

There is little evidence to determine to what extent public school teachers adopted Post's curriculum. At first it garnered little attention. One exception was Diane Roskies' *Teaching the Holocaust to Children* (1975) in which she offered a scathing critique of the small body of available research, literature, and curricula on the Holocaust up to that point—most of it intended for Jewish schools. She concluded that by 1975 Holocaust curricula "had gotten worse," "[it was] bland and full of clichés," and did not deal with "the essential nature of anti-Semitism." She wrote the "interest and fascination with the Holocaust among educators and young people in America is suspect," because they seemed to have no interest in the "heritage of those Jews who died."[11]

While Post's mini-course did include a lesson on Jewish life before the Holocaust, it did not escape Roskies' critical eye. She objected to Post's approach to the Holocaust as a case study, concluding, "The whole approach to the Holocaust as 'case study' in genocide can only diminish the uniqueness of the tragedy." What did she mean by the "uniqueness of the tragedy"? She was asserting that the Holocaust was a Jewish event that could only be understood in the context of Jewish history. Comparing the Holocaust to other genocides was ignoring this context and, therefore, historically inaccurate.

Roskies' complaint was buried in a rather obscure publication, but the issue of Holocaust uniqueness was a growing concern among Jewish scholars and educators. The controversy would explode in the autumn of 1977, when the New York City Board of Education recommended making the study of the Holocaust mandatory in all its schools and decided to develop further their Holocaust teaching guide. The new curriculum would be based on Post's, who "directed the project and provided extensive editorial assistance." The discussion over the proper way to represent the Holocaust in the curriculum and popular culture had been a heated topic for years. In fact, Jewish intellectuals had been arguing among themselves over this issue since the 1960s. But the controversy that unraveled in the pages of the *New York Times* in 1977 represented the first time the debate crossed into the public discourse in a way that would engage non-Jewish Americans.

The *New York Times* Debate

In June of 1974, three Brooklyn Assemblymen announced their plans to introduce legislation that would mandate the teaching of the Holocaust in

New York City schools. "At the very most," Dr. Seymour P. Lachman, president of the New York City Board of Education, explained, "students know that the Nazis murdered six million Jewish men, women and children. In my opinion this neglect must not continue." They need to deal, he continued, with "the holocaust as the essential trauma of the 20th century."[12] The proposed legislation was part of a continuing rise in Holocaust consciousness among Jewish and non-Jewish Americans.

As mentioned earlier, over the course of the 1970s, the Holocaust moved to the center of Jewish consciousness, and, therefore, to the center of the New York Jewish community. Scholarly symposiums on the Holocaust hosted by the Cathedral Church of St. John and the Institute of Contemporary Jewry of Hebrew University of Jerusalem were held in New York City. The conferences received extensive coverage in the *New York Times* and included notable participants such as Raul Hilberg, Emil Fackenheim, Yehuda Bauer, and Saul Friedlander.[13] At these conferences, the leading Holocaust scholars discussed and solidified their positions on the historical and metaphysical uniqueness of the event. At this point, Holocaust uniqueness was largely a concern among Jewish intellectuals, but it was beginning to receive attention from Christians who considered the role of anti-Semitism in the event. Author William Styron, who would later go on to pen the Holocaust novel *Sophie's Choice,* wrote a piece for the *New York Times* in June 1974 reflecting on his trip to Auschwitz. Styron challenged the "oddly self-lacerating assertion . . . that the Holocaust came about as the result of the anti-Semitism embedded in Christian doctrine." He pointed out that at "Auschwitz not only Jews perished but at least one million souls who were not Jews." The overemphasis on German anti-Semitism, he argued, "ignored the ecumenical nature of that evil."[14] It was the ecumenical nature of evil, encompassed in the term genocide, upon which Albert Post based his unit. The question of whether the Holocaust was a result of a specific form of anti-Semitism or a more ecumenical evil would be at the center of the controversy over the NYC curriculum.

In addition, it should be noted that during this time the city of Philadelphia had designed and implemented its own Holocaust curriculum. In 1975, the Jewish Community Relations Council hosted a pedagogical conference on "Teaching the Holocaust." The conference inspired Ezra Staples, Franklin Littell, and the Philadelphia school system to design and distribute its own Holocaust curriculum.[15] Mirroring the actions of the New York City Board of Education, in 1977, the Philadelphia school system moved to make the curriculum mandatory in all its ninth grade world history courses. The German-American Committee of Greater Philadelphia protested the 127-page curriculum guide that gave "the impression that

Nazis were the only ones who committed crimes against humanity and that the Jews were the only ones who suffered to any great extent." Rev. Hans S. Haug, head of the committee, suggested that the course reduce the amount of material on the Jews and add material on what he called genocides by the Soviet government, by Moslems and Hindus in India in the 1940s, and by the Communist government of Cambodia."[16]

Despite the potential for controversy, the New York City Board of Education, Division of Curriculum and Instruction, with the help of Albert Post worked for two years on the its own Holocaust curriculum, still titled *The Holocaust: A Case Study in Genocide*. The finished product was over four hundred pages and "was prepared in response to a wide demand by concerned citizens and educators to support curricula on the subject of genocide." The lessons aimed "to draw parallels between the forces and events which gave rise to Nazism, and contemporary events and issues which relate to the Holocaust." Like Post's curriculum, it was meant to be flexible enough for teachers to adopt it as a unit, a mini-course, or a semester elective. Like Post's curriculum, it also included large excerpts from Elie Wiesel's 1972 *New York Times* article as a rationale for teaching the event. The curriculum contained over a hundred excerpted readings from primary and secondary sources on prejudice, genocide, anti-Semitism, prewar Jewish life, persecution, and resistance. It also included suggestions for stimulating class discussions on the "Holocaust as a universal danger."[17] Initially introduced into a limited number of schools, the Board members announced that they hoped the curriculum would be mandated by the following year.

In the autumn of 1977, a brief article appeared in the *New York Times* on the new Holocaust curriculum that kicked off a debate that would rage for several weeks. The article reported on the objections to the study of the Holocaust by certain interest groups. George Pape, president of the German-American Committee of Greater New York, claimed that teaching the Holocaust would create "a bad atmosphere towards German-Americans in this country," and that "there was no real proof that the Holocaust had actually happened." Dr. M.T. Mehdi, "president of an Arab-American group," was even less diplomatic. He proclaimed that the curriculum was "an attempt by the Zionists to use the city educational system for their evil propaganda purposes," and asserted that if the Holocaust were to be mandated, then the City should add "the study of slavery and other acts of genocide." There was no evidence that either critic actually looked at the curriculum. The reporter Ari Goldman only included a one-sentence description of the curriculum and used the remaining space to outline the different controversies surrounding its implementation. If Goldman was hoping to incite a debate, he was quite successful.[18]

Five days later Paul Ronald wrote a letter to the editor in support of Holocaust education, but only "if the term Holocaust is not essentially limited to the murder of millions of Jews." He then listed the "multitudes of victims" who were murdered by "the Nazis and their allied scum" such as "Free Masons, Jehovah's Witnesses, Catholic and Protestant clergymen, gypsies, Socialist and Communist leaders and workers, French and Italian civilian hostages, Polish and Russian civilians." Three days later, the *New York Times* printed two more letters they had received. The first was by Isle Hoffman, head of the (German-American) Steuben Society Council's Education Committee, who protested the proposed Holocaust curriculum, assuring that it would "be divisive and serve no purpose other than to incite new atrocities." Hoffman asserted that it was time "to recognize the futility of accusing multi-ethnic Americans of the sins of ancestors in their county of origin." There was, in fact, some evidence of growing anti-German sentiment. In Philadelphia in 1978, after a Holocaust curriculum had been implemented, the German-American Citizen League reported that students with German surnames were called Nazis by their classmates.[19]

Nevertheless, the second letter that day was by Howard Marcus, a Holocaust survivor who had lost his whole family to the Nazis. He attacked George Pape's assertion that there was no proof the Holocaust actually happened and offered his own experience as refutation.[20] Holocaust denial had actually received an "official" endorsement the previous year by Arthur Butz, a professor of electrical engineering at Northwestern University. His book *The Hoax of the Twentieth Century* (1977), which cast the Holocaust as a Jewish conspiracy, had garnered a considerable amount of attention in the press. While it sought to disprove the Holocaust, it actually aided the cause of Holocaust education. Supporters wished to offset the lies and misinformation being disseminated by Holocaust deniers and looked to Holocaust education as a corrective. Despite his standing as a college professor, Butz's book was so detached from mainstream scholarship that being associated with him would translate into political suicide.[21] For this reason, George Pape felt obliged to soften his position on the Holocaust.

A month later, Pape wrote to the *New York Times*, claiming his Holocaust denial had been taken out of context. He wrote the article "appears to use excerpts from my off-the-cuff remarks to create controversy where none exists." He then presented his official position of the German role in the Holocaust, "there is no question in my mind that countless civilians were slaughtered for political purposes... among them Jews who were only killed for being Jewish... And certainly, not among the least of the political systems operating this deliberate slaughter was the Nazi regime."[22] Pape pointed out that there were "countless other" victims of the Nazis and that other countries participated in the destruction of the Jews besides Germany. While he certainly did not accept the full responsibility of Germans, his position effectively distanced him from the political volatility of Arthur Butz's Holocaust denial. Professor Butz's book brought attention to the issue of historical research and what kinds of conclusions should be accepted by the public. In response to Holocaust deniers such as Butz and a growing cadre of American neo-Nazis, the media cast Holocaust historians as defenders of Holocaust memory—a role they often reluctantly took. Israeli historian Yehuda Bauer was one scholar who was comfortable defending the Holocaust in public. He used the forum of the *New York Times* debate over the curriculum to express his official positions on the role of Germans, Jews, and non-Jews in the Holocaust.

Bauer was one of the leading Holocaust historians in the world, and author of several books. Drawing upon his research that would be published the following year in *The Holocaust in Historical Perspective*, he suggested "one of the problems of teaching the Holocaust to teenagers is that, unfortunately, there are gradations of evil." He then outlined the complicated position on the historical uniqueness of the Holocaust and its relation to the other groups persecuted by the Nazis. He explained how the murder of the European Jews was unique because "Jews could not escape—no apostasy, no identification with Nazism, no change in domicile within Nazi Europe helped at all. All persons were under a sentence of death who had committed the crime of having had Jewish grandparents." Bauer also offered a less-than-enthusiastic endorsement of Holocaust education as a means of preventing future genocides. "The Holocaust should be taught as an event of universal importance precisely because of its uniqueness," Bauer continued, "There is an off-chance that this might help in trying to prevent things from deteriorating again."

In his letter, Bauer used the term "unique" four times in reference to the Holocaust. While this was not the first time the term was used to describe the event, Bauer's letter represented its most conspicuous use to date. It may have been the first time non-Jewish readers were confronted with the historical "uniqueness" of the Holocaust.[23] The uniqueness of the Holocaust and its similarity to other events became a central issue to how the curriculum should be framed. The *New York Times* took an official position on October 28.

The editorial argued that "the annihilation of the European Jews should be a mandatory subject." It explained the nature of the controversy surrounding the proposed curriculum, which, at this point, "has been covered extensively in some schools, cursorily in some others and not at all in still others." The author asserted that "there are good reasons for emphasizing the fate of European Jews," but that it was also important for students to learn about "the famine that drove the Irish to these shores, the prejudice that Italians found when they arrived, the harassment of German Americans during World War I, the forced relocation of Japanese-Americans during World War II."[24] The writer was careful to list each major ethnic group of New York, perhaps, in an attempt to put the issue to rest. If this was the case, it did not work.

More letters were printed in the weeks that followed, both for and against the proposed curriculum. Walter J. Fellenz, a retired U.S. soldier who liberated Dachau in April 1945, gave his full endorsement to New York's proposed curriculum. "No, I am not a Jew," he wrote "I am but a soldier. And I congratulate the authorities in New York City... for plans to teach the Holocaust in their public school programs."[25] Philip Reiss, a history professor of German descent who grew up in a Jewish neighborhood, also supported Holocaust education. He offered his personal experience as refutation of George Pape's assertion that learning about the Holocaust would incite anti-German American sentiment, "I never experienced any rancor directed at me by Jewish neighbors." He suggested that Jews wanted the Holocaust to be taught "as a reminder—the lessons of which should be benefit to all people."[26] Those readers who had any direct or indirect connection to the Holocaust pledged their support for the mandatory curriculum.

On the other hand, retired teacher Eileen O'Connor suggested that teaching about past atrocities would only exacerbate ethnic tensions. "To arouse in every ethnic group," she worried, "hatred for the miseries of past generations, by instilling in the youth all such history, will create a multiplicity of civil wars worse than any we have ever known." O'Connor represented the assimilationist values of the previous generation, who looked to consensus and equality as the guiding American paradigm toward minorities. "As we embark on exploration in space," she wrote, "we should not think in terms of races of men or national groups, but of man as humanity." By the 1970s, the vigilant work of minority leaders had replaced this rosy view with a multicultural one.

Minority groups looked to their own ethnic past as a means of achieving cultural unity. Since their history looked considerably different from the "white" history appearing in textbooks that either ignored or distorted their past, they began to produce their own works, which concentrated on oppression, prejudice, and victimization. American minority groups, the Jews included, wanted to express how their past made their identities distinct. O'Connor felt that multiple histories would further divide the nation. "Why can we not leave history to the ages?" she asked, "I sincerely hope schools will drop such history of hate and will teach constructive subject matter." The following week, Nathan Belth wrote a reply to

Ms. O'Connor. Echoing the famous words of philosopher George Santayana (i.e., those who don't know their history are doomed to repeat it), he suggested that if the Holocaust is not taught "then we may indeed be doomed to repeat the past—hopelessly to the end of time."[27] In the end, Belth's position was the one that was more representative of New Yorkers.

The Holocaust and Genocide Education

The *New York Times* debate touched on a number of Holocaust-related issues. To what extent was the Holocaust historically unique, as compared to American slavery, the Irish Potato Famine, or other genocides? To what extent was the Jewish experience different from the other minorities who suffered under the Nazis such as Jehovah's Witnesses, homosexuals, and gypsies? To what degree was German anti-Semitism different from a more universal, "ecumenical" hatred? Would teaching the Holocaust assuage rising ethnic tensions, or exacerbate them? These questions were all concerned with how to place the history of one ethnic group into the context of others. In the emerging paradigm of multiple histories, how would one atrocity sit alongside the rest? Would such protests inevitably lead to what historian Peter Novick would later call "the Victimization Olympics?"[28]

For critics, New York City's Board of Education was sending the message that the Holocaust was worthy of a mandate, and thereby, more important than other cultural histories that would merely be covered in existing textbooks. The Holocaust was chosen because, according to Arnold Webb, executive director of the Division of Curriculum and Instruction, "the intent and scope of mass murder are unprecedented." Like Albert Post's original curriculum, the published NYC curricula ultimately took a contradictory position on the historical uniqueness of the Holocaust. The curriculum was still primarily concerned with the facts surrounding the murder of the European Jews (including a section on anti-Semitism), but it also suggested that "the Holocaust is one of several case studies that can be taught" such as "the experience of enslaved Blacks in the North and South America, Native-American Indians defeated in wars with the United States, Christian Armenians exterminated by the government of Turkey, . . . the murder of millions of Cambodians and the fratricidal conflicts in Northern Ireland." Its authors wrote, "the lessons [of the Holocaust] are universal" and suggested that "emphasis should be placed on concepts such as racism, ethnocentrism, prejudice scapegoating, [and] stereotyping." The balance between the unique and universal aspects of the event was uneasy.[29]

The *New York Times* debate also spoke to a number of curricular issues. To what degree should history be connected to contemporary concerns? To what extent did comparing historical events distort or misrepresent them? Can learning about a particular historical event really impact individual or social conscience? Can history be used to teach morality? If so, whose morality? Eileen O'Conner, the only educator in the debate, questioned the benefit of concentrating on how ethnic groups had been wronged in the past. She, along with certain Arab American and German American groups, thought that learning about the Holocaust would increase ethnic tensions. On the other hand, the NYC Board of Education and the majority of letter writers to the *New York Times* thought that learning about the Holocaust would increase student tolerance. In the years that followed, the anti-Holocaust education critique would be quelled, as a general consensus emerged that learning about the Holocaust was in some way beneficial for all students. Attention would shift toward exactly how to teach and frame the event.

The use of history in the schools is at the heart of the debate over Holocaust education. Should history be used to unite a culturally disparate people, as O'Connor suggested, or should it be used as an impetus for critical thinking and social amelioration? This unresolved tension had been present for a century, but the socio-curricular changes of the 1960s and 1970s added another level of complexity to the issue. John Dewey and the social studies reformers had argued for a curriculum centered on student needs and interests. But Dewey and his progressive contemporaries assumed that, at any one time, a monolithic set of student values would emerge around which to organize the content. Progressives presupposed an evolving consensus of American values to work toward. But the socio-curricular context of the 1970s emphasized multiple cultures and multiple values. There was no longer a consensus of values to work toward, nor was there a collective history to transmit. It was not enough, the multiculturalists argued, for students to study their own ethnic history; students also needed to study other American cultural histories as well, in order to empathize with different cultural perspectives.

This view was characterized by Rabbi I. Usher Kirchblum of the American Jewish Congress in 1974, who explained "ethnic-studies programs are not enough to deal with this issue. The problem with ethnic studies is that they are elective rather than compulsory. As a result, by and large, only black students enroll in black studies, only Jewish students take Jewish studies and so on."[30] Learning about the Holocaust would help students to empathize with historically oppressed groups. But these multiple histories overloaded an already bloated curriculum and raised the issue of whose histories would take precedence over others. There was not room for everything.

Teaching the Holocaust as history had curricular problems as well. On one end of the spectrum, it was clear that placing too much emphasis on Jewish victimization would belittle the suffering of other ethnic groups. On the other, attributing the Holocaust to an "ecumenical evil," as William Styron had suggested, would deemphasize the uniqueness of the Jewish experience under the Nazis. Such a universal approach, historian Yehuda Bauer argued, was historically inaccurate. The New York City curriculum made an awkward compromise by studying the Holocaust through the lens of genocide. The curriculum suggested that the Holocaust was an "unprecedented" example of genocide, of which the Jews were primary victims and for which anti-Semitism was a necessary precondition; but the curriculum also pointed out that there were also other Nazi victims and other examples, though not precedents, of genocides to which the Holocaust could be contrasted, but not compared. This approach satisfied no one completely, but at the time, it seemed like the best solution. As we saw in the last chapter, Jewish educators had been grappling with these very issues themselves. When New York City designed its curriculum, Jewish educators had not reached their own consensus position on the meaning of the Holocaust.

The *New York Times* debate demonstrated the cultural tensions that genocide education could spark. However, the issue of how to frame the Holocaust in a culturally acceptable way did not seem to be a central concern for Albert Post. Although his exact pedagogical motivations were unclear, he did not design his curriculum to appease Jewish critics, who had been urging for more inclusion of Jewish history since the early 1960s. Nor was his curriculum likely a direct response to Wiesel's article. So what did inspire Post to introduce his curriculum in 1973? As the next chapter will demonstrate the first designers of Holocaust education were driven by the curricular zeitgeist of affective learning. They hoped that the teaching of relevant material would excite and engage their students in history. They were reacting to a perceived educational need—a motivation completely missed by contemporary critics of Holocaust education.

Chapter 3

Affective Revolution and Holocaust Education

Learning about the Holocaust aligned with the overall goals of the comprehensive high school, an institution that was experiencing a great degree of financial and cultural stress in the 1970s. Secondary schools were designed to bring students from disparate social and ethnic groups together and to keep them engaged (and enrolled) with relevant material. Many secondary school teachers and administrators were realizing that these goals were often contradictory. Discussing slavery and racism openly with minority students could be counterproductive by exacerbating racial tension. On the other hand, discussing the history of rich white men could bore and antagonize students, who were understandably cynical of "official" history in the wake of Vietnam and Watergate. The topic of the Holocaust, however, provided a way to talk about discrimination and apathy indirectly. It provided a window into a whole slew of issues and events for students to explore safely in ways that made them feel subversive. Many asked why schools had ignored the Holocaust for so long. Why were Americans ignoring genocides in the present? As we shall see such ideas did not simply emerge from local classrooms, but instead were part of a larger movement away from traditional history toward areas of conflict and diversity known as the "affective revolution."

Engaging Student Values

Affective revolution is the term that social studies theorists Gerald Marker and Howard Mehlinger applied retrospectively to educators' intense interest during the late 1960s and 1970s in students' identity, morality, emotions, and values. These reforms lost momentum at the end of the 1970s when they were replaced by the "back-to basics" and standards movement.[1] Yet the short-lived affective revolution created an educational environment conducive to the introduction of the Holocaust as a topic of study. The researchers and educators involved in this movement were all either building upon or reviving theoretical ideas introduced by progressive educators, especially those of John Dewey.

In educational circles, as well as within the culture at large, the turbulent 1960s marked a turning point. The nation witnessed the assassinations of John F. Kennedy, Martin Luther King, Jr., and Robert F. Kennedy. The war in Vietnam continued to escalate, triggering protests, civil unrest, and the formation of a counterculture. Racial tensions resulted in violent riots in major cites such as Detroit, Los Angeles, and Washington, D.C. In this environment, many educational theorists began to challenge the idea that a democratic society was inherently moral and just. American and individual values began to be questioned. Social unrest seemed to peak in 1968, as students across the county demanded rapid curricular reform that would speak to their issues and concerns. "Relevance" became the educational buzzword of the day.[2]

The National Council of the Social Studies (NCSS), a body formed by history teachers, social scientists, and social studies professors in 1921, responded to these changes by including more material on racial issues in its curriculum and focusing its attention on student activism and identity. It devoted the entire April 1969 issue of *Social Education*, its flagship journal, to minority issues.[3] Accordingly, in 1971, NCSS rewrote its curriculum guidelines to reflect "the massive efforts to revise the social studies curriculum that have taken place in the last decade."[4] The guidelines demonstrated new emphasis on relevance and reaffirmed that the curriculum should be "directly related to the concerns of students"; it should "deal with the real social world."[5] A curriculum that spoke directly to students' issues accorded perfectly with the growing body of research on students' values and moral growth.

While the affective revolution seemed to have burst onto the scene at the end of the 1960s, its research foundation stretched back at least a decade. Several researchers had been working to reconcile the pedagogical issues of individual and group values in a democratic society by drawing upon—but revising—the work of progressive educators such as John Dewey. The roots of the affective revolution in the social studies curriculum can be traced to the work of Donald Oliver, James Shaver, and Fred Newmann in the late 1950s and early 1960s. Shaver and Newmann were doctoral students under Oliver at the Harvard Graduate School of

Education. Together they worked on the Harvard Social Studies Project, which was partially supported by federal funding made available after the launching of the Soviet satellite *Sputnik*. The project was intended to bring rigor to the social studies curriculum. These researchers produced dozens of instructional pamphlets on a variety of historic and civic issues, including Nazi Germany.

On the basis of their work with high-school students, Oliver, Shaver, and Newmann argued that the social studies should center on problem areas of relevant public issues and the resolution of values conflicts. While this idea had precedents in the era of progressive education, these researchers suggested a slightly different approach to morality. In their 1966 study, Oliver and Shaver rejected Dewey's relativistic, pragmatic approach to values on the grounds that it failed to deal with "extreme violations of important social values, which may be justified rationally."[6] They critiqued Dewey's view that actions can be guided by prediction of their possible long-term consequences. In a particularly apt example, the authors made reference to the destruction of the European Jews by questioning the appropriateness of debating "the reasons for and against Hitler's extermination of the Jews in central Europe in terms of possible consequences." They asked, "While there may be justifiable homicide, is there ever justifiable genocide?"

The authors argued that while the concept of truth in a pluralistic society cannot be defined "in such unequivocal terms that all will see it and grasp it in the same way," everyone generally can agree on the most basic level that the purpose of a democratic government is "to promote the dignity and worth of each individual who lives in the society."[7] The protection of human dignity was a common ground on which all parties could converge, even if their individual conceptions of this ideal differed. Oliver and Shaver promoted what they called the "jurisprudential approach," or the systematic analysis of public issues immediately affecting students' lives. Students would learn to distinguish facts from values and arrive at a compromise position. The focus on classroom deliberation seemed to be derived from Oliver's own teaching style. According to the *Harvard University Gazette*, he was known for conducting marathon discussion classes, lasting for hours or more. He was remembered by his students for his "confrontational discussion style."[8]

The "jurisprudential approach" was complicated but involved eight steps: abstracting general values from concrete situations, using general value concepts as dimensional constructs, identifying conflicts between values constructs, identifying a class of values-conflict situations, discovering or creating conflict situations that are analogous to the problem under consideration, working toward a general qualified position, testing the factual assumptions behind a qualified values position, and testing the relevance of statements. The jurisprudential approach to the social studies was not meant to replace the content of history, but to serve as a lens through which it could be viewed. It would also provide a guide for determining what sort of "content" should be selected. History, they argued, should be taught as a continuing conflict of values and tied to the immediate needs of society.

However, an even greater influence on the "affective revolution" was that of Harvard psychologist Lawrence Kohlberg, who addressed issues of morality from a cognitive point of view. Like Dewey, Kohlberg's theory emerged out of a critique of the two available approaches to moral education, what he called the romantic and the cultural transmission models. The former involved letting morality develop naturally. The latter involved imposing a preconceived "bag of virtues" onto a child. According to Kohlberg both theories did not consider the socially constructed nature of moral development, which included developing, "structures, internally organized wholes or systems...for the processing of information."[9] Both systems, he suggested, ultimately led to moral relativism. The romantic model sought to isolate and protect the student from the encroachment of others' values, leading to disparate and varied moral systems. On the other hand, the cultural transmission model merely passed on the values of politically correct majority culture, a set of ideals that inevitably evolved over time. Both were inherently relativistic; both were inadequate.

Kohlberg's theory of moral development drew upon the pragmatic philosophy of John Dewey and the research of Swiss psychologist Jean Piaget. As Kohlberg explained, Dewey viewed "the acquisition of morality as an active change in patterns of response to problematic social situations rather than the learning of culturally accepted rules."[10] Morality, according to Dewey and Kohlberg, is neither the internalization of established cultural values nor the unfolding of spontaneous impulses and emotions. Rather it is "justice, the reciprocity between the individual and others in his social environment."[11] This idea was developed further by Piaget, under whom Kohlberg studied. Piaget's empirical research confirmed Dewey's philosophical insights. Drawing upon his study of young children's subject–object relations, Piaget developed a theory that would eventually become to be known as "constructivism." He argued that children (subject) employed a particular mental structure, or schema, to process new information. When this structure no longer worked adequately to explain the physical world (object), they experienced disequilibrium. They were then forced to construct a new schema or mental paradigm to reconcile them selves with the dissonant information. This process involved an overall epistemological shift, which was tied to biological processes, but was also dependent upon social interaction (especially when the "object" became other people and their views).[12] As Kohlberg asserted, both Piaget and Dewey claimed "that mature thought emerges through a process of development... resulting from organism–environment interactions."[13] He employed these theories to the development of human morality.

Kohlberg's cognitive framework involved six distinct stages. His stages progressed gradually from an "obedience–punishment orientation," in which one simply followed the rules to avoid extrinsically imposed consequences, to a "conscience or principle orientation," in which one's actions arose from an intrinsically derived set of principles. More specifically, the moral stages of development included: (1) obedience and punishment orientation—egocentric deference to superior power or prestige, or a trouble-avoiding mindset; (2) naively egoistic orientation—right action is that which instrumentally satisfies the self's needs and occasionally those of others; (3) good-boy orientation—orientation to approval and to pleasing and helping others; (4) authority and social-order-maintaining orientation—orientation to "doing duty" and showing respect for authority and maintaining the given social order for its own sake; (5) contractual logistic orientation—recognition of an arbitrary element or starting point in rules or expectations for the sake of agreement; and (6) conscience or principle orientation—orientation not only to actually ordained social rules but to principles of choice involving appeal to logical universality and consistency. This framework, he explained, was composed of "basic moral concepts believed to be present in any society [that]... must be conceived of in terms of cognitive-structural changes... rather than in terms of the learning of cultural patterns."[14] In Kohlberg's case the term "affective" was not appropriate, because it implied that values were inherently relativistic. By tying moral development to the cognitive processes of the mind, Kohlberg argued that his stages were universal. However, these stages were actually thresholds, capacities that individuals could achieve. One did not necessarily perform at the highest level consistently, instead one often regressed to previous levels depending on the situation and context.

Kohlberg applied his theory in an educational context. Movement from one stage of moral development to the next was not inevitable. It involved "the communication of definitions of situations which elicit socially appropriate affect."[15] In other words, it required that students be confronted with hypothetical and real moral dilemmas and work them out through class discussion. Guided peer discussion, according to Kohlberg, would stimulate movement from one stage of moral reasoning to the next.[16] It was particularly important for individuals to consider the values of others and try to empathize with them. "Morality," he wrote, "is a natural product of a universal tendency toward empathy or role-taking... [and] a product of a universal concern for justice, for reciprocity or equality in the relation of one person to another."[17] Kohlberg's cognitive developmentalism provided the research-based conceptual framework that teachers needed to introduce class discussion of contemporary issues. Such discussions would not only be "relevant" to student concerns, but also grounded in the latest psychological research as well.

In the mid-1960s, Louis E. Raths developed a different approach to moral development—one that he called "values clarification." Also building on the work of Dewey, values clarification was "not concerned with the content of people's values, but [with] the process of valuing." As did Kohlberg's approach, values clarification offered a developmental framework composed of several subprocesses: prizing and cherishing, publicly affirming, choosing alternatives, choosing after consideration of consequences, choosing freely, acting, and acting with a pattern. But, unlike Kohlberg's approach, values-clarification left itself open to moral relativism and cultural imperatives. Kohlberg realized this and pointed out that values-clarification "does not attempt to go further than eliciting awareness of values." If this program were followed, Kohlberg alleged, "students [would] themselves become relativists, believing there is no right moral answer."[18] While researchers could not reach a consensus on how to engage and develop students' moral values most effectively, they became increasingly concerned in the late 1960s and early 1970s that historical education had neglected student values.[19] New materials that did not address the moral aspects of history were irresponsible and incomplete; the social context of the times demanded a curriculum that addressed students' affective domain.

The affective revolution resulted from students' desires for a relevant curriculum that would provide them with direction in turbulent times. While the idea of organizing the curriculum around student concerns was first put forward in the progressive era, questions of racial injustice, victimization, and civic disobedience complicated the issue. Values could no longer be centered on the progressives' assimilationist paradigm. By the late 1960s, American values seemed irreconcilably divided. In educational circles, the minority equal rights movement soon morphed into multiculturalism. For educational multiculturalists, the confluence of American values was undesirable and oppressive. Instead, they argued, the values differences of ethnic minorities should be recognized and celebrated.[20] Oliver, Shaver, Newmann, Kohlberg, and Raths supplied a theoretical foundation for considering such values-conflicts in the classroom. They agreed that morality was not something imposed upon students, but rather something that emerged from within them through social interaction with their peers.

In their writings, these researchers cited the Holocaust as an extreme example of moral corruption and as a rebuttal to extreme moral relativism.[21] This social and curricular context, crossed with a rising interest in the Holocaust, inspired a group of teachers to introduce the topic into the public schools in the mid-1970s. In learning about the Holocaust, these teachers hoped, students would explore the relationship between the individual and society in a way that would be relevant to their own lives. The extremity of the Holocaust helped teachers find a foundation of moral consensus on which to build their value-laden discussions; everyone in education, regardless of ethnicity, could agree that Nazism was evil and that the Jews were innocent victims. Thus, teachers in both religious and public school settings used the Holocaust to activate the moral reasoning of their students.

Playing Holocaust

In 1969, a student demonstration erupted outside the General Assembly of Jewish Federations in Boston. The young protesters denounced the current state of Jewish education and demanded a curriculum that was more relevant to their lives. One of the conference attendees was Rabbi Raymond Zwerin from Denver, Colorado. He recalled how the students were decrying "the lack of significant Jewish educational material for the Jewish school." The Rabbi translated the student protests into a mandate to develop new educational materials that would speak more directly to Jewish student concerns. When he returned to Denver, he worked with Audrey Friedman Marcus, a sixth grade religious school teacher, to form the Alternatives in Religious Education (ARE) Publishing Company, dedicated to designing "cutting edge" educational materials for interested teachers.[22]

Despite the substantial literature on the Holocaust in Jewish journals, many Jewish educators were still not covering the event by the early 1970s, due to the decentralized nature of Jewish education. As late as 1974, Jewish leaders were deploring the lack of knowledge and understanding among Jewish youth about the Holocaust.[23] Part of the reason for this apparent lack of interest by Jewish youth, Rabbi Raymond Zwerin suggested, was the "pathetic" state of available Jewish educational material. "Until the late 1960s," he explained, "religious texts had the look of the late 1940s... history textbooks [consisted] of black and white stuff, written for their grandparents,... by a generation of assimilationists." Rabbi Zwerin was ordained in 1964 at the Hebrew Union College-Jewish Institute of Religion in

Cincinnati and later founded the Temple Sinai in Denver, Colorado. He became highly active in local Jewish community and helped found the Babi Yar Park Holocaust Memorial in Denver. Audrey Friedman Marcus had been active for years in the Jewish educational community by leading workshops and designing modern curricular materials.

When Zwerin and Marcus created ARE in 1973, they aimed to design educational materials that were "succinct, interesting... educationally sound and self-motivating" by integrating some of the "new techniques that came out" in Christian and public schools, such as "the beginnings of values-clarification."[24] Their first project together was a series of pamphlets on Jewish holidays, traditions, and community. Soon ARE designed dozens of lessons, covering the entire range of Jewish educational topics. The lessons were specifically aimed at part-time teachers "with more passion for the job than knowledge," who didn't have the time to develop their own materials. Two of ARE's earliest projects were minicourses that specifically addressed the Holocaust in creative ways, *The Holocaust: A Study in Values* (hereafter referred to as *A Study in Values*) and *Gestapo*: *A Learning Experience About the Holocaust*. Both these minicourses took an affective curricular approach to teaching the event.[25]

The authors designed *A Study in Values* to explore the moral complexities of the Holocaust period. The curriculum was centered on fictional interview transcripts of German bystanders and perpetrators who lived through the Holocaust years. The course objectives stated that the participant would be able to "express in his or her own words some of the attitudes and positions vis-à-vis moral issues which were held by many of the German people." Students were asked to consider each case study individually and, after group deliberation, decide on the character's guilt.

For example, one case included Has Brenner, a supervisor of an electro-welding shop that oversaw several Jewish laborers. Brenner was primarily concerned with meeting production quotas, a goal that relied upon the intensive, unpaid work of Jews. The character denied any knowledge of Jews being killed by the Nazis. "I personally did not see any of them die. Some collapsed on the job and were carried away. I do not know whether they died or not." He also denied any feelings of guilt. "Why should I?" he explained, "There was nothing I could do about it. If the Jews did not receive pay for their work, that was their problem. If they did not work hard, that was our problem, because our bonuses would be less." After considering the facts of the case in regards to international and Jewish law, students were expected to reach a verdict for each fictional participant.[26]

The curriculum also encouraged students to compare the moral dilemmas of both Germans and Jews who lived during the Nazi period to contemporary events. In fact, each case study was accompanied by discussion questions that addressed "contemporary/universal issues." The questions for the "Case against Hans Brenner" suggested that students consider issues of discrimination (sex, age, minority), the right to demand certain benefits for past inequities (affirmative action, quota systems), and unfair labor practices (slavery, child labor, sweat shops). Students' "attitudes and values" would be "brought out through discussion." The teacher's guide also provided extensive historical background for each case study, which the teacher was expected to share.

The authors designed *A Study in Values* to transmit the historical facts of the Holocaust period by using the fictional case studies as representations of the larger Nazi bureaucracy. The curriculum avoided large generalizations about the victims and perpetrators, thus, avoiding the simplistic narrative of good versus evil. When viewed on the individual level, it was clear that many Germans acted in the muddy area between these extremes. Even in the Holocaust, ideals of morality, according to the Rabbi Zwerin and Marcus, were more complicated than many Jewish textbooks had implied. Still, the curriculum authors were not suggesting that students adopt a view of moral relativism. They suggested that students consider international and Jewish law to judge the guilt of these fictional characters. The objectives asked students to "employ some of the moral dictates of the Jewish tradition in determining the guilt or innocence of representative individuals who lived during the time of the Nazi era." This affective approach would not only help student to "clarify" their values, but also make the events of the past relevant to the moral ambiguities of the present. Students were encouraged to apply the moral dictates they learned from the Holocaust to contemporary events.[27]

The *A Study in Values* curriculum was designed to stand alone as students' only exposure to the Holocaust, but the authors suggested that the minicourse be followed by the *Gestapo* game. The *Gestapo* game was a Holocaust simulation that covered the years 1933–45 in a chronological manner. The game could accommodate up to a thousand players, depending on how many boards and value markers were purchased. Each player received a "*Gestapo* Value-Board" containing the following categories: house, community, life, income/job, pride, family, religion, and civil liberties. Students were given three value markers for each category. As the leader read a card describing an event from the Holocaust, students had to risk one of their value markers on each turn. Depending on what appeared on the card, players would either lose or keep their value marker. For example, one card read "Mass deportation of Jews begins. As yet there are no killing centers, so large masses of Jews are rounded up and placed in already existing crowded ghettoes." Those players who risked their community, house, pride, or family markers would have lost it on that turn.

The leader's cards progressed chronologically and accurately through the incremental events of the Holocaust. The cards were color-coded into three periods: white for the period 1933–38, during which "losses in income/job, civil liberties, pride and community were sustained by the Jews, due to the passage of debilitating laws"; blue for 1939–41, "the ghettoization and deportation phase"; and pink for 1942–45, "the Final Solution." Players were given the opportunity to trade cards and occasionally escape opportunities would arise. When a player lost all of his/her life cards, he/she was out—"a victim of Gestapo." The resulting percentage of survivors was designed to approximate that of the actual Holocaust. Those that survived the *Gestapo* game would be those who took great risks early in the game, because "Sometimes you have to throw yourself into it wholeheartedly," Rabbi Zwerin explained, "to stop a process." In all aspects the game was designed to provide a realistic representation of how the Holocaust transpired and the value decisions that the victims had to make.[28]

Gestapo and *A Study in Values* were both designed to engage secondary-level students with relevant questions and to challenge their value systems. Both of these objectives aligned with the goals of the affective revolution. About the same time that Zwerin designed his curricula in Colorado and Albert Post began implementing his in New York City, teachers in New Jersey and Massachusetts were also implementing Holocaust units in their classrooms. While these teachers also drew upon the general ideas of the affective revolution, as we shall see, their rationales drew more directly upon the theories and research of Lawrence Kohlberg.

Society on Trial

Roselle Chartock of Great Barrington, Massachusetts, designed one of the first Holocaust units intended for use in a secondary public school in 1973. An edited form of her curriculum, timed to coincide with the airing of NBC's *Holocaust* miniseries (see next chapter) appeared in NCSS's journal *Social Education* in April 1978. The unit approached the topic from a psychological perspective aimed at probing the nature of humankind. It contained an abundance of visual and written sources that attempted to transmit the emotional trauma of the victims and the moral dilemmas of the perpetrators. The unit shifted back and forth between the historical facts of the Holocaust and investigations into the nature of man.

The story of Chartock's curriculum began in the fall of 1972 when Jack Spenser, chairman of the Social Studies Department at Monument Mountain

Regional High School, applied to the National Conference of Christians and Jews for a grant to develop a Holocaust unit. Spencer and Chartock received an award of $2,135 and designed the unit over the summer of 1973. The designers approached the Anti-Defamation League of B'nai B'rith (ADL) in the fall of 1974, asking them to publish the curriculum.[29] After several revisions by Chartock, the unit was published in 1978 as *The Holocaust Years: Society on Trial.* The alliance with the ADL resulted in the inclusion of the unit in *Social Education.* The ADL's program director Theodore Freedman wrote an introduction for the Holocaust-themed issue of the publication. Freedman suggested that teaching the Holocaust was "one way to defuse prejudice" by making it "intellectually indefensible" and that the Holocaust could "also help students deal with the unanswered questions of blind obedience to authority, reawakened to some degree by Vietnam and Watergate."[30] The ADL would financially support a number of Holocaust education efforts by non-Jewish public schools teachers in the years to come, but it would only do so through contacts with specific teachers who were already experimenting with Holocaust education. They did not have an active agenda of recruiting teachers or using their political and financial influence to push the topic into the public schools.

Chartock's abridged unit was included in the periodical along with articles by Elie Wiesel and Raul Hilberg.[31] The curriculum employed an affective pedagogical approach to the Holocaust, dealing with the "complexities of man's behavior under various conditions" through an investigation into individual values and morals. It viewed the Holocaust not as a historical or cultural problem, but as an ecumenical problem of humankind. The curriculum was influenced by Kohlberg, whose work, Chartock wrote, was "useful in the instruction of a problem-oriented interdisciplinary curriculum, particularly the Holocaust curriculum." Since "moral dilemmas are actually the central core of this curriculum," Chartock explained, "the moral reasoning process becomes a necessity in dealing with this material." Although she did not accept wholeheartedly the validity of Kohlberg's theory of moral stages, Chartock agreed with its "basic premise" that "controversy and conflict, which [are] part of any problem-oriented curriculum [are] to be accommodated or resolved by reasoned reflection—cognitive and moral—and by conversation rather than force or coercion."[32]

Chartock's approach de-emphasized the historical particularities of German society and the specific circumstances of the European Jews. Instead, she attributed the Holocaust to the "the undesirable aspects of man's relationship with his fellow man . . . a product of the part of human nature that could be sparked in many places and times, wherever and whenever certain conditions are present." Thus, an underlying assumption of her unit was that Holocaust-like persecution could break out in America at any time. In the tradition of the social studies, Chartock stressed that the content of the course should be connected to the lives of students. "The courage and morality of a society," she asserted, "are constantly on trial."[33] Students had to learn how to be decision-makers rather than ignorant cogs in the machine of society. They had to aspire to reach the higher levels of Kohlberg's moral stages in order to be able to act with an individuality that transcended cultural norms. Learning about the Holocaust, Chartock suggested, could help achieve this goal.

The interdisciplinary aspect was crucial to Chartock's approach. In order to elicit the affective responses she wanted from her students, she had to present "the feel of the conditions of dehumanizing... the emotional impact." A purely historical account of the Holocaust failed to do this. For her purposes, it was necessary to incorporate visual and first-person reflections on the event. "The novel and film," she explained, "provide... the feelings for the reader that the factual content and theories of the social science cannot."[34] She hoped to trigger emotional responses in her students by exposing them to first-hand accounts and actual images of the Holocaust, as opposed to "make believe" scenarios.

Her unit began with a screening of the thirty-three-minute French film *Night and Fog,* which showed graphic images from the Holocaust such as, in her words, "naked men and women with heads shaven, gaunt, hollow-eyed faces staring through barbed wire, showers that spewed gas, ovens full of ashes of human bones, [and] piled-high bodies being bulldozed into open pits." The film, Chartock explained, introduced her students to "this period in history in as dramatic and... as *authentic* a way as possible" (her italics). Exposure to actual images of the Holocaust, she suggested, would impart to the students a sense of the reality of the event that could not be fully replicated through the cold, objective words of historians. The film shocked the students and aroused an intense interest. After showing the film, Chartock encouraged her students to write down "some of their feelings" about the film. Her technique was in line with the educational literature on affective learning, which suggested that students would be more interested in the material if their values, emotions, and morals were triggered and tested. The images sparked their desire to learn. The students, Chartock explained, wanted to know how those images could be real, and to discover what kind of people could have created such massive destruction and held such disrespect for human life. She reported that students left the classroom asking "Why? Why? Why?"[35]

The unit followed the showing of *Night and Fog* with various readings, including Elie Wiesel's *Night*, that further described the events of the Holocaust. Chartock related that her students—even those who disliked reading—responded energetically and with curiosity to the reading assignments. Next, the unit shifted focus to more universal themes. Students read works by John Locke, Thomas Hobbes, and Niccolò Machiavelli as a way to spark discussions on the nature of mankind. They considered the question of whether humans were inherently good or evil. The readings were difficult, and many students required Chartock's help. But, she reported, their intense interest spurred them on. The curriculum then engaged students in a role-playing activity in which they took on the identities of German bystanders during the Holocaust period. This was consistent with the central mode of development in Kohlberg's theory of cognitive developmentalism, which relied on role-playing as a means of inspiring empathy in students for those who held views different from their own. Role-playing exercises, according to Kohlberg, enabled students to progress to the next developmental moral stage.[36] "As a result of role-playing," Chartock explained, "students were able to identify their own values as well as the values of people different from themselves." She reported that "it was as if a time machine had propelled them back into the 1940s." The role-playing enabled her students to focus on the moral and emotional responsibilities of those Germans and Jews who were alive during that period and on how their decisions related to similar situations in their lives.[37]

Chartock wanted her students to understand the complexity of acting morally in difficult situations. She assured her students that while the Holocaust had occurred over thirty years before, society was faced with similar dilemmas in the 1970s. An underlying assumption of this approach was that morality was an ahistorical phenomenon; Kohlberg's research on cognitive developmentalism supported this. Chartock's role-playing exercise suggested that the moral and value options of her students were identical to those of the Germans in the 1930s; historical and cultural context was insignificant. Her students could identify with the Germans because, in essence, they were all human. They all had the same inherent propensity for good and evil. Because this approach ignored the historical particularities of the social, cultural, and intellectual milieu that surrounded the Germans, it left students with an unrealistic view of that time.

The unit also did not provide any moral standards against which to judge the Germans. While Kohlberg's cognitive developmentalism provided a conceptual foundation, Chartock's unit did not instruct students in his theories. As a result, students could adopt a stance of moral relativism—the view that the Germans' moral responsibility should be judged against each individual German's value system. In fact, Kohlberg designed his framework as a counter to this kind of moral relativism, an approach he associated with "values-clarification." Chartock asserted that moral responsibility was a matter of interpretation: "Students discovered that there was no single answer to the question of 'Who was responsible for the Holocaust?'" She explained, "there were alternative interpretations that depended on a person's values and frame of reference."[38] Chartock hoped that through the study of the Holocaust, and specifically of the moral ambiguities that faced Germans, her students would arrive at a more refined understanding of their own values. This would enable them to face the moral ambiguities of contemporary society, which was the major objective of the affective curricular approach to history.

Chartock was not only one of the first educators to design a Holocaust curriculum for use in the public schools, but she was also a leader in helping other educators design their own. As the following case studies will demonstrate, Chartock's curriculum served as a model for other interested teachers such as Richard Flaim and Edwin Reynolds of New Jersey and William Parsons and Margot Stern Strom of Brookline, Massachusetts, all of whom would go on to become even more influential in Holocaust education than Chartock herself. Through her cooperation with the ADL, Chartock hosted Holocaust education workshops for dozens of teachers and thereby provided the initial impetus for a grassroots educational movement.

A Search for Conscience

Richard Flaim and his colleagues at Vineland High School and Edwin Reynolds of Teaneck, New Jersey, began teaching the Holocaust experimentally in the mid-1970s. In the fall of 1975, the ADL and the New Jersey Education Association encouraged Flaim and Reynolds to examine Chartock's work. With Chartock's encouragement, the Vineland teachers introduced their own elective course entitled "The Conscience of Man," which centered on the events of the Holocaust as a way of investigating the nature of humankind. While Chartock's unit was one of the first implemented in a public school, the Vineland course would be one of the first full-semester classes on the Holocaust in the entire country.

Vineland, New Jersey, in 1973 was a community of fifty thousand people with an ethnically and socioeconomically diverse population. Richard Flaim, head of the Social Studies Department, reveled in the multiple cultures of his community. He reflected that it was "almost like a laboratory for social studies education... a good place to do interesting things with kids to prepare them for a multiethnic society and world." In 1973, Flaim and his colleagues sensed that their social studies curriculum was no longer adequate for the turbulent times. They decided to "make it more relevant and more important for the future of our kids" by focusing on the "moral and ethical issues that emanate from historical events." He recalled that his students were overwhelmed by the civil rights movement and the resulting black nationalism, the women's liberation movement, and the antiwar protests. His students "were really struggling to find their own way, to know what was right to do... kids had all kinds of questions." His department made a conscious effort "to help students examine [these] moral and ethical issues." This need demanded a reconceptualization of the curriculum and a review of the available materials.[39]

Flaim had been certified to teach in 1960, but as department head, he worked hard to keep up with the latest social studies research. He was familiar with the work of social studies theorists such as Fred Newmann and Edwin Fenton, but he was particularly "influenced by the work of Lawrence Kohlberg," whose theories of cognitive-moral development he passed along to the teachers in his department. He hoped that Kohlberg's framework could serve as the foundation for a new, more relevant curriculum. "Once we had a goal that related to cognitive-moral development," he recalled, "we then started looking at pieces of history with a new lens." Harry Furman, one of the younger teachers in the department and a former student of Flaim's, suggested the Holocaust as a potential topic. Furman's parents were Holocaust survivors, and he pointed out that in Vineland there were two hundred survivors "about whom our children knew nothing." Flaim and Furman agreed that the Holocaust was an important and neglected topic, and that it could serve as a conceptual and factual basis for launching discussions on moral and ethical issues facing society. The teachers began to work on a semester-long course entitled "Conscience of Man" that would concentrate on the Holocaust.[40]

In addition to receiving suggestions and support from Roselle Chartock and Jack Spencer of Great Barrington, Flaim and Furman also contacted Leatrice Rabinsky, a teacher from Cleveland Heights, Ohio, who had been experimenting with teaching Holocaust literature in her school. Flaim, however, expressed the opinion that his course would have to be different from Chartock's. Although neither community contained a large percentage of Jews, Vineland was far more diverse than Great Barrington, which was predominantly white. "We developed our program to meet the needs of our community," Flaim explained.[41] The curricular freedom that characterized the 1970s allowed these teachers to experiment with a variety of topics, approaches, and materials without the hindrances of state testing and top-down impositions of reform. This curricular freedom was a precondition for the launching of Holocaust education in American public schools. By the early 1980s, this inconsistency would come under attack from various quarters, including that of certain educators who blamed the new social studies and the affective revolution for the state of uncertainty in the social studies curriculum. But without this spirit of experimentalism, the Holocaust education movement might not have taken off.

By 1976, the materials were ready and Vineland High School offered an elective course on the Holocaust, taught by Harry Furman. Despite his reputation as a demanding teacher, students signed up for the course with enthusiasm. The Holocaust was a subject of great interest, Furman explained, since "kids are curious about human nature; they want to talk to each other about these things. Our class gives them the chance to do that." Positively received by students, the class also positively affected students' perspectives and behavior. Flaim recalled that "a young Hispanic boy stood up and said that the reason the course was valuable to him was that it had made him look at a Jewish classmate in a totally new way." The boy explained that before the course, he "didn't know anything about Jews... now I know that I have to look at everyone as an individual." Flaim remarked that hearing this young person talk about his revelation was one of the most moving experiences of his career.[42]

Across the state in Teaneck, Edwin Reynolds and his colleagues in the social studies department were teaching a similar course. Teaneck had a larger Jewish population than Vineland; Jewish students made up one-third of the total student body there. In 1976, the *New York Times* sent a reporter to sit in on the class, and an article appeared in June describing the discussion that ensued. According to the article, many of the students had not heard of the Holocaust before they signed up for the course. They were shocked by the images that they saw and the content that they learned. The reporter wrote that students were deeply engaged in discussions of the event. Considering whether a Holocaust could occur in America, one student commented: "If things got bad enough in this country, the people would rather see six million persons killed than have their own family killed." Another student commented: "Look at Watergate. We allowed it to happen and now we have already forgotten." In their discussion, the students pondered the moral responsibility of the Germans. "It was the entire society," one student concluded; "people didn't care. That's what allowed it to happen. You have to make moral judgments." Another interjected: "The main thing is for people to care enough to have moral judgment."[43] The teachers in both Vineland and Teaneck were getting the responses that they had desired. Students were engaged in the materials and were making connections to contemporary events on their own. They were coming to understand the consequences of inaction and the importance of moral responsibility.

By the spring of 1978, the teachers in Vineland and Teaneck had been in contact with each other to exchange ideas. In addition, Flaim had worked with the Cumberland County Jewish Federation in Vineland and the local and New York ADL to contact Holocaust survivors and to share his ideas and resources. The ADL had initiated a fund-raising drive the year before to develop and house Holocaust educational materials in New York. But until the NBC *Holocaust* miniseries aired in April, the League had not intended to distribute materials across the state. In response to parents' and educators' demands after the series was broadcast, the New Jersey Board of Education approached the teachers in Vineland and Teaneck to request that they develop a resource guide for teachers throughout the state. The Board provided funding for Flaim and Reynolds to begin work in the summer of 1978 on a state-endorsed curriculum.[44] The New Jersey ADL coordinated work on the curriculum and invited hundreds of organizational leaders to participate in the design process at six statewide meetings. After several drafts and subsequent piloting in schools, the ADL published the New Jersey curriculum in 1983. It was called *The Holocaust and Genocide: A Search for Conscience.*[45] The primary editors of the anthology and curriculum guide were Richard Flaim and Edwin Reynolds. Harry Furman, Ken Tubertini, and John Chupak were contributing editors. Furman was the only one of the five teachers who was Jewish. The curriculum approached the Holocaust from a more social–scientific perspective than did Chartock's. The New Jersey teachers used the event to study the nature of man.

The teacher's guide began with an introduction by Governor Thomas Kean in which he introduced his inclusive definition of the Holocaust as an event that "engulfed not only Jews, but all those unfortunate enough to live under the cloud of Nazi domination." By 1983, the issue of Holocaust uniqueness could no longer be circumvented or ignored. Kean addressed the issue directly, pointing out that while the Holocaust was not unprecedented, it represented the first time that "the best resources of the state [had] been legally dedicated to mass murder." One notable precedent that Kean identified was the "wholesale destruction of the Armenians by the Turks," a recognition that would later incite objections by Turkish Americans.[46] The governor suggested that the Holocaust contained lessons for every American, young and old, about the role of the individual in the modern state. Governor Kean explained that the Holocaust was particularly relevant for the students of New Jersey, the second most diverse state in the Union.[47]

The curriculum began with a statement of rationale that reflected Kohlberg's theory of cognitive developmentalism. The study of the Holocaust, the authors asserted, would contribute to the "moral as well as cognitive development of the students" through the discussion of "some of the most crucial moral and ethical questions which have faced, and continue to face, the human race." The unit approached the Holocaust through a lens of rational detachment. Students were provided with a conceptual and theoretical foundation on the basis of which they could engage some of the moral ambiguities surrounding the Holocaust.

Unlike Chartock's unit, which sought to trigger emotional responses in students through film, literature, and role-playing exercises, the New Jersey curriculum encouraged students to arrive at "conclusions about human nature"—specifically, conclusions that could be applied to other events. And while Chartock never provided her students with a moral framework, the New Jersey curriculum suggested that teachers introduce "The Kohlberg Scale of Moral Reasoning...as a means of comprehending the capacity of individuals for blind participation in actions which result in suffering." Appropriately, the first two units of the New Jersey teaching guide were on "The Nature of Human Behavior" and "Views of Prejudice and Genocide." Thus, students developed a moral and conceptual framework before they learned about the historical particularities of the Holocaust. The guiding question of the course materials was how mankind could have allowed the Holocaust to happen. The designers felt that this question could best be answered through an investigation of human nature; the historical facts of the German situation would constitute one of many tools in the inquiry process.

True to spirit of the affective revolution, the New Jersey course connected the content to the present needs of adolescents. The course concluded with a unit on the "Related Issues of Moral Responsibility," in which students discussed the relevance of the Holocaust to atrocities past and present, and considered the possibility of a Holocaust in American society. This was the part of the curriculum that the *New York Times* reporter observed. To make the course resonate with students, the designers included dozens of contemporary learning tools in the curriculum. For example, the unit on the nature of human behavior suggested that teachers introduce the lyrics of rock songs such as Billy Joel's "The Stranger," The Rolling Stones' "Sympathy for the Devil," and King Crimson's "Twenty-First Century Schizoid Man." The presentation of the Holocaust was designed to have the greatest possible effect on students. In this sense, its purpose was not only to transmit the facts of the Holocaust, but also to transform the attitudes of future citizens—a central goal of the social studies in general. This aim would be even more clearly discernible in the Massachusetts curriculum *Facing History and Ourselves* (hereafter referred to as *Facing History*).

Facing History and Ourselves

The *Facing History* curriculum was designed by two teachers in Brookline, Massachusetts, who met at a Holocaust conference sponsored by the New England ADL in 1974. The organizers had chosen to hold the conference at Bentley College, rather than at the more obvious location of Brandeis University (which had a large Jewish population), because they wanted to present the Holocaust as "a human problem, not a Jewish one." While all seventh- through twelfth-grade teachers in the Brookline area were invited, only three attended, including Margot Stern Strom and William Parsons; Stern Strom's parents were Jewish, and Parsons was the son of a Methodist minister. The conference included lectures on various aspects of the Holocaust and on the moral responsibility to remember.

Parsons and Stern Strom were disturbed that though they both had graduate degrees in history, they had learned very little about the event. They decided to design their own Holocaust curriculum as a way to combat societal ignorance. In the summer of 1974, Parsons and Stern Strom received a grant from Brookline Public Schools to develop lessons on the Holocaust. They knew that in addition to transmitting the facts, they had to create a program that would "link a particular history to universal questions, those timely yet timeless questions that resonate with every generation."[48] They had to design a curriculum that would connect the events of the past to the dilemmas of the present.

Like Roselle Chartock and the New Jersey teachers, Parsons and Stern Strom were intellectual, creative, and ambitious in their approach to the course. However, they were unusually qualified to design and teach this type of material. Parsons had studied under Fred Newmann at the University of Wisconsin at Madison. In the 1960s, Newmann had worked with Shaver and Oliver in the Harvard Social Studies Project, where they researched and published material on values conflicts and public controversy in the classroom. While Newmann was at Harvard, he and Oliver had designed a unit on Nazi Germany that made reference to the destruction of the European Jews.[49] After Newmann left Harvard for Wisconsin, he continued to promote Harvard's "jurisprudential approach" to public issues—which he updated in 1970 in *Clarifying Public Controversy: An Approach to Teaching Social Studies.*

In this work, he stressed that "teachers should not try to persuade students to take particular views on issues, but help them develop a style of justification." Newmann argued that affective and cognitive domains were impossible to separate, and so rather than teaching students the "right" answers about areas of value/political consensus, educators should concentrate on areas of value/political conflict. They should teach students how to defend their values in a rational manner. Ultimately, Newmann insisted, students should be encouraged to act upon their beliefs, and he suggested that "students receive educational credit for participation in causes of their choice."[50] Parsons was greatly influenced by Newmann, from whom he learned that relating the facts of history was not enough; teachers had to inspire their students to become participating citizens—to take action.[51] Parsons' exposure to Newmann prepared him to work with issues of moral and values conflict. As it turned out, his future partner Stern Strom was equally inspired.

Stern Strom grew up in Memphis, Tennessee, during the period of segregation—a time in which systemic and individual racism were unquestioned. Her teachers, she recalled, "did not trust us with the complexities of history... its legacies of prejudice and discrimination." But such moral dilemmas were not overlooked by her parents, who were both Jewish and openly discussed such issues with her. Her mother taught Sunday school and was very active in the local Christian community. Her father routinely took Margot to temple. From an early age, she was confronted with the realities of both racism and anti-Semitism and discussed these issues with her parents. When Stern Strom moved to Massachusetts, she was further influenced by Catholic priest Father Robert Bullock, whom she considered her "most significant teacher." Without him, she reflected, "there would be no Facing History and Ourselves." While her cultural background equipped her with the moral foundation to engage the topic of the Holocaust, her educational background was equally strong.

Stern Strom studied the theories of cognitive developmentalism at Harvard Graduate School of Education, where she was enrolled in Lawrence Kohlberg's "moral development program." She had first-hand knowledge of the research that had influenced so many of the pioneering Holocaust educators. In the years to come, *Facing History* would benefit from the cooperation and support of some of the leaders in cognitive and educational psychology, such as Carol Gilligan and Howard Gardner of Harvard. Parsons and Stern Strom combined their intellectual resources to design a curriculum that accorded with the latest social studies and cognitive theories. Their approach, therefore, was not based on anecdotal evidence or vague rationales, but on an emerging research base. They began teaching their program in the fall of 1976.[52]

In the 1970s, Brookline was segregated along racial and socioeconomic lines, and this resulted in what Parsons referred to as the "ghettoization" of the city. Stern Strom's school was mostly middle-class with a substantial Jewish population. Parsons described his community as "blue-collar, mostly Black and Hispanic."[53] Despite the different demographics, the response to the curriculum was positive in both schools. One parent related to Stern Strom that in no other course had her daughter "been exposed to real dilemmas as complex and challenging... [or] had she been inspired to use the whole of her spiritual, moral and intellectual resources to solve a problem." After the successful implementation of their curriculum, Parsons and Stern Strom applied for and received a federal Title IV-B (Federal Elementary and Secondary Education Act) grant for schools with underprivileged children. They used this grant to establish the Facing History and Ourselves Foundation and to pilot their program throughout the Brookline school district.

By 1978, Parsons and Stern Strom were employed full time by the Foundation. They worked on improving their curriculum, hosting awareness sessions, running workshops, and supporting teachers in the field in implementing the program. The curriculum received another boost in 1980 when the U.S. Department of Education recognized *Facing History* as an "exemplary model education program" and added it to its National Diffusion Network for use in schools across the nation. In 1982, the curriculum was published by the Foundation for national distribution.

The curriculum approached the Holocaust through a social scientific framework, which was reflected in its subtitle, "The Holocaust and Human Behavior." The designers wanted students to understand what had happened and why it had happened, not how they themselves might have reacted in similar situations. Aside from the inquiry into the nature of human behavior, the unifying theme of the program was genocide. The educators used the Ottoman Armenians and the European Jews as the main examples. They wanted to relate the full story of these two major events, which had been ignored in textbooks. The curriculum's introduction included a fascinating note from a textbook company speculating on why the Holocaust had received only superficial coverage to date. "The topic probably is regional in localities where there is strong Jewish community;" the note's author explained, "I suspect that there are a number of things already available for free from various Jewish groups." In addition, the Holocaust represented what he called "a very high risk situation, with students sharing experiences and attitudes." The discussion of values in the classroom, he worried, extended beyond the public school's mandate and could potentially incite parental protest. Like the New Jersey teachers, Stern Strom and Parsons found the general lack of knowledge about the Holocaust appalling. Their curriculum aimed to rectify the situation by fully covering the facts.[54]

The *Facing History* curriculum introduced a new element into Holocaust education: a concern about nuclear war. The earlier curricula had connected the Holocaust to contemporary concerns such as Watergate, civil rights, and the Vietnam War, but by 1982, while these issues had not necessarily been resolved, they no longer commanded the attention that they had in the mid-1970s. Accordingly, the authors of *Facing History* connected the Holocaust to a concern that they considered more pressing: nuclear proliferation. Indeed, in the 1950s and 1960s, the term holocaust was used commonly in reference to a nuclear war—more so than in reference to the destruction of the European Jews.[55] Fears of nuclear holocaust had subsided with a thawing of the Cold War culture in the late 1960s and the 1970s, but in the 1980s President Ronald Reagan moved nuclear proliferation to the forefront of America's political and social consciousness.

The *Facing History* curriculum outlined the similarity between the global silence surrounding the Holocaust during World War II and the global acquiescence to an arms buildup with the potential to lead to a nuclear war in the present; both events, it posited, required a majority of passive bystanders too apathetic or frightened to intervene. "The problems involved in confronting information about contemporary nuclear issues and the potential for nuclear war," the authors explained, "are remarkably similar to those described in discussion about confrontation with the Holocaust." The authors suggested that Americans were apprehensive about discussing controversial or disturbing material, especially in schools. But, the authors argued, this type of reluctance was a precondition for future mass atrocities. Their curriculum aimed to instill in students "the power of the individual to make decisions which affect not only the protection of his or her neighbor, but the survival of the world." Students were encouraged to apply what they learned in class to protest against nuclear war. In accordance with Fred Newmann's work, the curriculum encouraged a "sense of community and the impulse to be more socially active."[56]

The curriculum also tackled issues of moral and cognitive growth that were the central focus of Kohlberg's work. The methodology was designed, its authors explained, "to encourage students to understand more than one perspective in a dilemma, to place themselves in the position of another person." To avoid moral relativism, students were introduced directly to Kohlberg's stages of moral development and the international definition of genocide as tools for evaluating the behavior of the historical actors that they encountered. Students confronted the Holocaust from a behaviorist perspective, examining specific historical events to see what they revealed about human nature. As in the New Jersey program, students began with units on "Society and the Individual" and "Anti-Semitism," which were designed to help them form a conceptual framework before they confronted the actual events of the Holocaust. The units that followed covered "German History: World War I to II," "Nazi Philosophy and Policy,"

"Preparing for Obedience," "Victims of Tyranny," "The Holocaust," "Who Knew? Individuals, Groups, and Nations," and "Judgment." One notable difference from the earlier curricula was in the sequencing of the content.

In *Facing History*, the "forgotten genocide" of the Armenians was introduced after the Holocaust had already been covered as a way to consider whether "we learn from past experiences." Thus, students were confronted with successive layers of content meant to complicate their assumptions about human behavior. The final chapter connected the material to contemporary society and asked what lessons had been learned. Here, the authors returned to their theme of nuclear proliferation as the next potential Holocaust. They warned that while the study of "the Holocaust does share certain basic principles with a study of potential nuclear holocaust, it should not be used to draw too strong parallels." Even so, the authors concluded the textbook with a reading on nuclear war. Students are implicitly instructed to take an active role in protesting nuclear weapons, lest another Holocaust break out.[57]

In 1982—the same year in which *Facing History* was published—a study sponsored by the National Jewish Resource Center on Holocaust curricula was completed. It evaluated the effectiveness of four major curricula then in use: a Philadelphia curriculum; a curriculum from Great Neck, Long Island; the New York City curriculum; and *Facing History*. In this comparative study, the researchers evaluated the different approaches and concluded that *Facing History* had a "social science/behavioral orientation," whereas the others had a slightly more "historical focus." The study related that, of the four curricula, *Facing History* scored lowest as an "emotional experience" but highest for holding students' interest and for increasing awareness of other groups and individuals. Overall, the researchers found the Brookline curriculum to be the best of the four.

The researchers explained that the *Facing History* students had learned a great deal about human behavior and were able to use the historical facts of the event to make informed judgments. "They [the Germans] were so upset," one Brookline student explained, "because they didn't have jobs and it was easy for Hitler to blame the Jews and for others to follow." Another student was less empathetic, explaining that the Holocaust happened "because people were stupid and did not understand or see that Hitler was gaining a lot of power. They believed him because they wanted to." These class discussions, one student commented, "make you think about your life and other people's lives." This formal evaluation enthusiastically endorsed Holocaust education as a whole, and specifically praised *Facing History*. The social scientific approach, the study concluded, seemed to engage students' interest and transmit the historical content more effectively than the emotional affective approach.[58]

Overall, Zwerin and Marcus' *Gestapo: A Learning Experience about the Holocaust* and *The Holocaust Study in Values*, Chartock's *Society on Trial,* and *The Holocaust and Genocide: A Search for Conscience,* and *Facing History* curricula each shared a few commonalities that corresponded with the underlying assumptions of the affective revolution. First, material on the Holocaust was selected for and organized around its ability to engage students, rather than its ability to reflect historical accuracy and context. This approach was manifest above all in the curricular focus on genocide rather than on the Holocaust as a historical event. In addition, the curricula included emotive elements that would engender in students empathy for the victims and bystanders, and potentially complicate their thinking about the inherent tendencies of humankind.

Second, students were encouraged to connect the events of the Holocaust to contemporary events and issues, including Vietnam, Watergate, nuclear proliferation, and racism in American society. Such activities, the designers argued, would help students navigate values conflicts in the present and empower them to prevent genocides in the future. Finally, these curricula applied other disciplines such as philosophy, psychology, and sociology to the "problem" of the Holocaust. Thus, the various disciplines were employed as curricular resources rather than curriculum itself. The commonalities shared by these curricula represent the underlying precepts of the social studies approach to history. Ever since the term was coined in 1916, the social studies has been a vague concept, but educators have consistently defined it in opposition to the traditional, lecture-based, "bag of virtues" model described by Kohlberg.

From these curricula we can also see the two main contexts from which Holocaust education emerged. In the years to come the two contexts would continue to perpetuate the movement. The first context, represented by New York City and Teaneck, New Jersey, were areas with substantial Jewish populations. The Holocaust was presented as having direct cultural relevance for students. Although connections were often made to other genocides and current issues, these were largely approached through the lens of the Jewish experience under the Nazis. Contrastingly, the other context, represented by Vineland, New Jersey, and Brookline, Massachusetts, were areas with high non-white minority populations. The Holocaust had indirect relevance for these students, who were experiencing prejudice and discrimination in the present. The Holocaust was approached through an investigation into human nature with the Jewish experience just one of many analogous examples. Both of these approaches had roots in the affective revolution, which called for cultural relevance and controversy.

Chapter 4

Watching and Defining the Holocaust

By the late 1970s, the topic of the Holocaust had been gradually gaining more attention by mainstream Americans. The historiography had continued to develop at an exponential rate. The mandating of Holocaust education in cities such as New York and Philadelphia had launched public debates on the topic, and numerous colleges and universities were offering Holocaust courses. Still, interest in the Holocaust was largely limited to urban and suburban areas in the Northeast with large Jewish populations. This would all change in 1978, the year in which the Holocaust entered the American national consciousness. In April of that year, neo-Nazis planned a march through the Chicago suburb of Skokie, Illinois, a community with an estimated seven thousand Holocaust survivors. Newspapers reveled in the freedom of speech issue surrounding the controversy and gave an unprecedented amount of coverage to Holocaust victims and their neo-Nazi nemeses. In response, thousands of non-Jewish residents from across the country pledged their support for the survivors. That same year, President Carter announced plans to begin work on a national Holocaust memorial, an initiative that would eventually result in the opening of the U.S. Holocaust Memorial Museum in 1993.[1] But the single most important event in disseminating Holocaust consciousness across America was the airing of the NBC miniseries *Holocaust* in 1978.

The Holocaust on TV

The *Holocaust* program elevated the event from a regional concern to a national one, and launched debates over the meaning and position of the

Holocaust in American life. The series was accompanied by the distribution of educational materials by numerous organizations. Accordingly, the National Council for the Social Studies (NCSS) capitalized on the new interest by dedicating their entire April 1978 issue of *Social Education* to the Holocaust. Recall that the issue included an abridged version of Roselle Chartock's eight-week curriculum, *Society on Trial.*

In April of 1978, the NBC *Holocaust* miniseries aired over four nights and was watched by an estimated 120 million Americans.[2] The docudrama followed over the course of a decade the lives of two fictional families, one of German Jews and the other of German Gentiles with strong Nazi ties. Members of these two fictional families participated in nearly every major event of the Holocaust including the Nuremburg Laws, *Kristallnact,* the Wansee Conference, the Warsaw ghetto uprising, Buchenwald, and Auschwitz. The series ended on an uplifting note when one member of the Jewish family survives and joins the resistance.

The Anti-Defamation League (ADL) speculated that the series communicated more information about the Holocaust to more Americans in four days than the previous thirty years combined. Likewise, the miniseries represented the first time that thousands of teachers across the country were confronted with the event. Joyce Arlene Witt, an Illinois teacher who would go on to design her own Holocaust unit, reflected, "My work in Holocaust education began in 1978 ... [when] a made for TV mini-series titled *Holocaust* aired on commercial television bringing, for the first time, the subject of the Holocaust to a large public audience."[3] New Jersey teacher Richard Flaim explained, "What [the show] did for Holocaust education was unbelievable ... it created an instantaneous awareness of the event."[4] The miniseries was so important to the spread of Holocaust awareness that when President Carter announced plans for a Holocaust Museum in September 1979, the Holocaust Commission spokesman felt obliged to mention that "the commission is not a spin-off of the movie."[5]

The popularity of the series was reinforced by an enormous amount of coordinating political, commercial, and educational activity that helped promote the show before its airing and ensure its discussion afterward. For example, the mayor of Washington, D.C., and the governor of Illinois both proclaimed the week following the series "Holocaust Memorial Week" in their respective constituencies. Within weeks of the airing, Gerald Green's paperback adaptation of the show had sold a million copies, and in response to the new demand for Holocaust literature, publishers began to push their own Holocaust-related books. Backlisted titles such as Lucy S. Dawidowicz's *The War against the Jews* and William Shirer's *The Rise and Fall of the Third Reich* also received renewed publicity campaigns and increased sales. The show demonstrated the enormous commercial potential of the Holocaust, an aspect of the event that would continue to vex Jewish Americans for years to come. Nonetheless, the *Holocaust* production was an enormous financial risk and NBC did all it could to ensure its success.

The social climate of the time was receptive to dramatic stories of ethnic persecution. The success of the African American drama *Roots* the year before demonstrated that American viewers would be receptive to historical dramas related to American minorities. NBC executives had anticipated a similar reaction to their show, but they knew that they needed to enlist the support of particular groups who could potentially object to its content. NBC hired Rabbi Marc Tanenbaum of the American Jewish Committee as a consultant. As a result, the series received his full endorsement. Prior to its airing, NBC executives had also arranged an advanced screening of the series for over forty clergymen and organizational leaders, who also responded positively.[6] In further cooperation with the American Jewish Committee, NBC designed and distributed an accompanying education guide to the series. They anticipated that a demand for such material would emerge as a result of the series. So did other interested groups.

Numerous Jewish and secular groups capitalized on the exposure that the miniseries brought to the Holocaust by distributing their own information on the event. Their efforts were supported by the fact that the Holocaust had been an emerging topic in certain local and Jewish schools for several years prior. For example, NCSS and the ADL had cosponsored a conference on Holocaust education in New York City in October of 1977, six months prior to the showing of *Holocaust*. When the miniseries aired, there was already a substantial amount of educational material available.[7] This allowed the educational and Jewish groups to produce high-quality material quickly in support of the *Holocaust* miniseries. A five-part study aid was prepared and distributed to schools by the National Jewish Interagency Project—a group of fifteen Jewish organizations that did not include the ADL.[8] The ADL joined with NCSS to prepare its own sixteen-page guide called *The Record*, which was designed to look like a mini-newspaper tracing the events of the Holocaust from 1933 to 1945.

The cleverly designed *Record* presented its information in newspaper-like headlines and columns. It included photographs, Nazi documents, an article by Elie Wiesel, contemporary *New York Times* Holocaust accounts, excerpts from secondary sources, a thorough Holocaust chronology, and a "Glossary of Terms."[9] The Jewish Community Council of Greater Washington sent *The Record* to fifteen thousand students and members of the Interfaith Conference in just Washington, D.C. alone.[10] Overall, the ADL would disseminate approximately ten million copies of this guide through distribution deals with local newspapers. Despite additional

Holocaust-related educational materials by the National Council of Churches, the Cultural Information Service, and the American Federation of Teachers, demand for Holocaust material by schools and religious groups could not be met.[11] The scope and intensity of this educational campaign, organized to coordinate with a specific event in pop culture, was unprecedented.

Many considered the greater exposure that the miniseries brought to the Holocaust a positive development. Religious leaders, writers, and educators praised the intentions and execution of the show. Presbyterian pastor Rev. William Harter proclaimed, "This is one of the most ambitious efforts in the history of the medium." Executive director of the National Urban League said, "We have much to learn from *Holocaust* and about the need for human solidarity" and William Safire wrote, "NBC-TV performed an enormous public service."[12] Others were more skeptical of the softening distinction between education and entertainment, between informing for the sake of public good and informing for the sake of profit. *New York Times* writer John O'Connor wrote, "for the overwhelming majority of broadcasting executives, the bottom line of profit is paramount...despite noble intensions on the artistic side of the project, the process is inevitably reduced to marketing." He accused the network of involving educators and religious groups in a national educational campaign to ensure the popularity of the miniseries, while unwittingly enlisting their endorsement.

The miniseries introduced the perception by some that the Holocaust was becoming too popular—an idea that would have been inconceivable just a few years before. Certain Jews wondered whether this new attention to the Holocaust could have detrimental effects on its representation and reception. The most apprehensive critic of the growing popularity of the event was Elie Wiesel. As early as 1972, Wiesel commented that the Holocaust was being "discussed freely, perhaps too freely and too much."[13] The increased attention that the *Holocaust* miniseries brought confirmed his worst fears.

On the first day the miniseries aired, the *New York Times* published a scathing review by Elie Wiesel, now a professor at Boston University. He wrote: "this TV series will show what some survivors have been trying to say for years...And yet something is wrong. Something? No, everything. Untrue, offensive, cheap: as a TV production, the film is an insult to those who perished and those who survived...It transforms the ontological event into soap-opera. Whatever the intensions, the result is shocking." He used the forum not only to attack what he called the trivialization of the Holocaust, but also to reinforce his view of the Holocaust as a metaphysically unique event. "The Holocaust is...not just another event," he exclaimed, "This treats the Holocaust as if it were just another event...Auschwitz cannot be explained nor can it be visualized. Whether culmination or aberration of history, the Holocaust transcends history." He concluded that the miniseries had done more damage to Holocaust memory than good.

As knowledge of the Holocaust grew over the course of the 1970s, so did the status of Elie Wiesel, who would often be asked to endorse a variety of Holocaust-related events. Nonetheless, his views on the Holocaust were not representative of mainstream Americans, or even mainstream Jews. Wiesel had been an early public proponent of Holocaust memorialization, remembrance, and education. Therefore, the authors of the New York City curriculum included his 1972 article as a rationale for their unit. Likewise, the ADL's *Record* included a transcript of the address he delivered at the NCSS-sponsored conference in October 1977. His writings were beautifully crafted and powerful, and by 1978 his memoir *Night* had become a classic. He had become the representative voice of the Holocaust survivor in America (and the world). So, it was not surprising that Holocaust educators often included his quotations in their work. But, these educators often ignored or misunderstood the specificity in which he viewed the Holocaust. Not only was the Holocaust definitionally and historically unique, but, for Wiesel, the Holocaust was an unanswerable question, "the ultimate event, the ultimate mystery, never to be comprehended or transmitted." In his review of NBC's *Holocaust*, he explained, "Only those who were there know what it was; the others will never know." Wiesel suggested a specific approach to the Holocaust as an unanswerable question—a question not even the survivors themselves could fully comprehend.[14] He felt that the integrity of the event's mystery must be preserved even when teaching about it. He offered a specific teaching style.

Wiesel's teaching philosophy emerged from his upbringing in the Hasidic tradition of Judaism in Europe. His largest influence was his maternal grandfather, an old Hasid who would share inspirational tales and songs of Hasidic deeds of the past.[15] Wiesel's memoirs are filled with affectionate recollections of "his favorite teacher," his grandfather, whose morality tales would greatly influence his own writing and teaching.[16] From his grandfather, Wiesel learned to employ a particular teaching technique, one that uses stories to create cognitive dissonance in the minds of the listeners. The listener is confronted with moral dilemmas, but not provided with the resolution or answer. In this manner, storytelling is teaching, and listening is learning. But listening is an active pursuit for Wiesel; it demands the participation of the student. "To listen to a story," he explains "is to play a part in it, to take sides, to say yes or no, to move one way or the other."[17]

Wiesel has adopted his grandfather's teaching approach for his own instruction at the college level at City College of New York and Boston University. For Wiesel teaching is to relate a story in an interactive manner—to have the responses and questions of the student shape the telling of the tale. "I like to know what the students want to know," Wiesel reflected, "they give me the lecture through their curiosity, participation, and questions."[18] Teaching is a cooperative meditation with the students, not a transmission of facts. It should trigger an affective and cognitive response, which will spark further questioning and uncertainty. Central to Wiesel's pedagogical approach is the use of silence. It is in silence that the learning takes place. Silence has been a central theme in the testimonies of many Holocaust survivors; it represents the unfruitful search for meaning in their experience.[19] This silence is the central theme in much of Wiesel's written work as well as in his teaching approach. "I think we have tried to transmit, to communicate certain words of that era," Wiesel explained, "but not the silence of that era."[20]

When teaching the Holocaust, Wiesel has exclusively applied this approach. He represents the Holocaust as an unresolved problem—a story without any meaning or resolution. For Wiesel the Holocaust can only be viewed as an infinite series of unanswerable questions that no one can or should even attempt to answer. He explained, "I know that as teachers we are called upon to transmit some certainties . . . But no certainty is eternal; only the quest is. The quest is human, all the rest is commentary."[21] Therefore, understanding or knowing the Holocaust can never be achieved. Its comprehension will always remain out of intellectual and emotional reach. The only appropriate response to the Holocaust is further questioning or silence.

To provide an example of Wiesel's pedagogical approach to the Holocaust in a way that relates directly to the secondary curriculum, I turn to an article that appeared in the April 1978 issue of *Social Education*, the flagship journal for the NCSS. Wiesel's article entitled "Then and Now: The Experiences of a Teacher" was a transcript of an address he delivered the year before at an invitational conference in New York cosponsored by the ADL and the NCSS. The article was constructed around a series of unanswerable questions, in which Wiesel combines his own horrific recollections from the concentration camps with meditations on his inability to represent or teach his lived experience adequately. Wiesel writes:

> I confess I do not know how to teach these matters . . . How do you teach events that defy knowledge, experiences that go beyond imagination? How do you tell children, big and small, that society could lose its mind and start

> murdering is own soul and future? How do you unveil horrors without offering at the same time some measure of hope?[22]

The Holocaust cannot be understood, Wiesel asserted, because the mere physical facts that historians uncover can never adequately explain its dimension. Wiesel does not deny that the historical facts of the Holocaust exist, but rather claims that historians can never explain the "how" or "why" that link these facts. Any historical attempt to explain the actions of the perpetrators normalizes their actions and denies the metaphysical significance of the event. For Wiesel the facts must stand by themselves, engulfed in an unending series of questions. The following excerpt demonstrates how the specific facts of the Holocaust only serve as a foundation for furthering questioning, but not understanding. Wiesel writes:

> But how can I, a teacher, explain to my students so many things related to the Holocaust? How can I explain to them the indifference of so many nations and so many leaders to so many Jews? How [do you] explain to students that it was Switzerland, humanitarian Switzerland, that suggested that Germany stamp Jewish passports with the distinctive "J" so that they could be refused visas everywhere, not just Switzerland?...How [do you] explain the *St. Louis* affair in which everybody followed the journey of the ship filled with refugees, which was not allowed to land in the United States?[23]

For Wiesel the facts only provide the ability to ask more specific questions—questions that can never be answered. Wiesel asserted that one should consult the historical "sources," but that these sources will not provide "consolation."[24] The sources will only engender more questions. Wiesel concluded his article with uncertainty on how teachers should relate the facts and stories of the Holocaust. He ultimately suggested his own approach, which was to teach about the slaughtered children. "When I teach these matters," he wrote, "I teach the children...because those children became philosophers, they became theologians, they became historians, they became poets."[25] The murdered children represent the ultimate example of the irrationality and senselessness of the Holocaust.

The final words of the article explain this point, "What can I tell you as a teacher that teaches? It is more than a matter of communicating knowledge. Whoever engages in the field of teaching the Holocaust becomes a missionary, a messenger."[26] He encouraged educators to teach the Holocaust, but not to explain it; they must merely pass on its sacred mystery as messengers. Earlier in the article he reflected on his experience in the concentration camp, writing of "that extraordinary experience between the killer and his victim; something happened there, something theological, metaphysical, something transhistorical and historical. I cannot comprehend them."[27] This excerpt represents the most direct statement of his view of the Holocaust as a sacred mystery. It is fascinating that Wiesel would assert the theological significance of the Holocaust to public school social studies educators, who certainly would not be able to present the Holocaust to their students as a religious lesson. Nonetheless, his suggested approach was consistent with his view on Holocaust uniqueness, which was fully explained in his review of the *Holocaust* miniseries. Even though adherence to his particularist views on the Holocaust made teaching the event in a universalized or even secular manner impossible, Wiesel's popularity and stature continued to grow in the years following the miniseries. That same year President Carter would ask him to head his Commission on the Holocaust.

The miniseries not only helped to expose the specifics of Wiesel's philosophy, but it also sparked further debates over the meaning of the Holocaust in public life and in the public schools. Most of these issues had been raised before. Did the Jews die like sheep to the slaughter? Is the Holocaust historically unique? Have there been other genocides since the Nazis? Should children learn about the Holocaust? Would increased interest in the Holocaust fuel anti-German-American sentiment? Articles, editorials, and letters to the editor, which took positions on these issues, filled the pages of the *New York Times* in the weeks that followed.[28] These same issues remain a continual source of contention.

Holocaust Uniqueness

Holocaust uniqueness has been and still is a central concern for scholars of the Holocaust. Although the first teachers of the Holocaust felt obligated to address the issue, their approach to the event was driven by pedagogical concerns, not necessarily theoretical or political ones. Up to now I have been employing the modifiers definitional, historical, and/or metaphysical to describe the different elements of the uniqueness claim. I offer this trichotomous framework as a critique of previous discussions on Holocaust uniqueness and pedagogy, which tend to conflate the various elements of the uniqueness claim or place the conflicting views along a single continuum.[29] These distinctions emerged out of the intellectual discourse of the past thirty years. In the next section, I will temporarily step out the historical narrative to explore the different elements of the uniqueness claim. Since the *Holocaust* miniseries brought many of these issues to the attention of mainstream Americans, this seems like an appropriate place for this discussion.

The uniqueness of the Holocaust has been at the center of the debate about how to represent the event properly in memorial and popular culture. The term "unique" has been used with a variety of meanings. For some unique has meant that the Holocaust is an event that should be treated like no other, requiring a specific set of rules and restraints. For others unique has meant that the Holocaust was specifically a Jewish event and can only be understood in Jewish terms. It was this view that led Elie Wiesel to protest, "they are stealing the Holocaust from us."[30] The "they" to which Wiesel referred were the other minority groups who were also persecuted and murdered by the Nazis in World War II. These groups were seeking their own inclusion and access to the moral capital of the Holocaust. Comments like Wiesel's demonstrate how the representation of the Holocaust in America has proved to be controversial on cultural and historical grounds. The uniqueness claim has been used in protest over so many Holocaust issues that, at times, it is difficult to decipher its true meaning. In this section I will unravel the different uses of this term.

I suggest that the uniqueness claim has three elements: metaphysical, historical, and definitional. When someone like Elie Wiesel refers to the Holocaust as unique, he is asserting all three elements of the claim. But, on the other hand, when someone like historian Lucy Dawidowicz suggests that the Holocaust is unique, she is referring only to the final two elements. The metaphysical uniqueness claim stirred up the most controversy for those Americans, Jewish and non-Jewish, who have attempted to portray the Holocaust in popular media or memorial.

As mentioned in chapter one, those who assert the metaphysical uniqueness of the Holocaust are often referred to as particularists. They have suggested that the Holocaust exists as a sacred mystery with a spiritual significance that can never be fully grasped or comprehended. According to Alan Mintz, particularist*s* approach the Holocaust as a "radical rupture in human history that goes well beyond notions of uniqueness... [into] a dimension of tragedy beyond comparisons and analogies... [and that] ... any cultural refractions of the Holocaust are often antithetical to its memory."[31] This view often renders the Holocaust as a mystical event that must be approached with reverence. Any encroachment on this reverence is considered a betrayal of the Holocaust victims. Emil Fackenheim voiced this view as early as 1970 at a conference in Jerusalem: "A Jew knows about memory and uniqueness. He knows that the unique crime of the Nazi Holocaust must never be forgotten—and, above all, that the rescuing for memory even a single innocent tear is a *holy task*" (italics in the original).[32] In 1977, at a conference in San Jose, Fackenheim would argue that denying this type of uniqueness would "insult" the dead.[33]

Naturally, particularists object to any fictionalized portrayal of the Holocaust in film or literature by those who have not directly experienced the event. Artistically, they believe the Holocaust should be approached only in nonfigurative language, and any representations should be void of the author/artist's persona interjecting between the event and representation. "To use special effects and gimmicks to describe the indescribable," Elie Wiesel wrote in his NBC *Holocaust* review, "is to me morally objectionable."[34] For Wiesel any artistic attempt to depict the Holocaust in popular culture can only trivialize it.

Those who assert the metaphysical uniqueness of the Holocaust are offended when the Holocaust is ever compared to any other historical mass atrocities or moral transgressions against mankind. They think the Holocaust should not be embedded in a history of genocide, but rather framed in a specific Jewish narrative or meta-narrative. They suggest that the Holocaust was not the result of a universal racism or bigotry, but rather the result of a specific form of European and German anti-Semitism. Daniel Jonah Goldhagen argued in his international best-selling *Hitler's Willing Executioners* (1996) that the Holocaust was merely a result of the Nazis unleashing this pent-up German anti-Semitism, a thesis that has been attacked by many scholars as too simplistic.[35]

Ontologically and epistemologically, particularists feel that the Holocaust can never be represented accurately or even fathomed by those who have not experienced it first hand. Straight historical accounts often fail to access what Saul Friedlander identifies as "deep memory." Deep memory is "that which remains essentially inarticulable and unrepresentable, that which continues to exist as unresolved trauma just beyond the reach of meaning" and void of any redemptive quality.[36] It is this deep memory that can certainly not be reached by artistic representations, but also cannot be accessed by traditional historical representations. Such views have sparked debates on whether the Holocaust requires a specific set of rules for its historical study. Here it is important to make a distinction between those extreme particularists, such as Wiesel, and those more moderate ones such as Friedlander. The latter considers issues of how to incorporate the testimonies of survivors (including their metaphysical interpretations) into an overall secular historical framework, whereas Wiesel would consider any secular historical framework inadequate. Traditionally, historians consider testimonies of an event (especially emotionally laden ones) based on memory less reliable than contemporaneous documentation. For the sake of clarity, it may be useful to consider Friedlander as a proponent of epistemological uniqueness, but not necessarily metaphysical uniqueness.[37]

Scholars asserting the historical uniqueness of the Holocaust also suggest that it was a singular historical event, the likes of which has never before existed. But these scholars will not go so far as to assert that the Holocaust is transhistorical, holy, or mystical. Instead, the event can and must be depicted in historical, secular terms.[38] Nonetheless, the Jewish experience in the Holocaust is historically unprecedented. In other words, even if the Holocaust can somehow be represented (a claim particularists deny), it should not be compared to other historical events such as the Turkish murder of Armenians during World War I, the Soviet atrocities under Stalin, nor, as was particularly relevant in the 1970s, the Vietnamese massacre by Americans at My Lai. For example, historian Deborah Lipstadt exclaims, "To suggest that disastrous U.S. policies in Vietnam...were the equivalent of genocide barely demands response."[39] Such comparisons may not be offensive, like the particularists assert, but are historically inaccurate.

Historians have devoted pages of research to assert the historical uniqueness claim. In *The Holocaust in Historical Context,* Steven Katz argues the case for the Holocaust being dissimilar in essential and significant characteristics from any other acts of group atrocity.[40] Emil Fackenheim states the case for the historical uniqueness of the Holocaust within the framework of genocide, a term that he concedes can also be applied to the Armenians, but "as a case of the class: intended, planned, and largely successful extermination, [the Holocaust] is without precedent and, thus far at least, without sequel."[41] He attributes its uniqueness to the scholastically precise definition of the victims [the Jews], the judicial procedures procuring their [German] rightness, the technical apparatus for human annihilation, and "most importantly, a veritable army of murderers."[42] The historical uniqueness doesn't necessarily lie in the victims, since one should not compare degrees of suffering, but rather in the perpetrators. "The Germans were unique enough," writes Avishai Margalit and Gabriel Motzkin, "because, more radically than anyone else in the last millennia, they denied the idea of a common humanity both theoretically and practically. They embodied this denial of humanity in the way in which they fused humiliation and extermination in their ridding the world of Jews."[43] Such classification works on the assumption that the Holocaust was not just another atrocity but the most extreme in the history of man, or as historian Yehuda Bauer argues, "there are gradations of evil...the Holocaust thus appears as an extreme and unique case."[44]

Of course, not all Jewish historians have subscribed to the historical uniqueness argument. Peter Novick, author of *The Holocaust in American Life,* offers a critical perspective on how certain American Jews have used the Holocaust for political purposes. He writes, "the very idea of uniqueness is fatuous, since any event—a war, a revolution, a genocide—will have significant features that it shares with events to which it might be compared as well as features that differentiate it from others."[45] Novick is convinced that the Holocaust can be represented just like any other historical event. In her book *Reading the Holocaust* (1999), Inga Clendinnen has taken more a subtle approach to critiquing Holocaust uniqueness. Her book aims to dispel the "Gorgon effect—the sickening of imagination and curiosity and the draining of will which afflicts so many of us when we try to look squarely at the persons and processes implicated in the Holocaust."[46] To overcome this effect, she suggests that "it is not enough to loathe the perpetrator and to pity the victim, . . . We must try to understand them both."[47] Indeed anyone who partakes in the writing of history and moves beyond the mere chronicling of events is making an attempt to understand it.

Finally, there has been a great deal of public debate surrounding the definitional uniqueness of the Holocaust. Does the Holocaust refer solely to the murder of Jews, or does the term include the other victims of Nazi persecution such as Gypsies (Roma), homosexuals, Russian POWs, Poles and other Eastern Europeans? Lucy Dawidowicz has distinguished the "special case" of the Jewish experience from that of other Nazi victims. She points out that compared to other groups the European Jews lost two-thirds of their population, and that the "murder of the Jews and destruction of communal existence were, in contrast, ends in themselves, ultimate goals to which the National Socialist state had dedicated itself."[48] She points out that destroying the Jews was actually counterproductive to the strategic gains of the Nazis.

Scholars frequently describe and respond to Holocaust uniqueness in their consideration of different elements of the event. Most of them respectfully refute the metaphysical uniqueness aspect. The same has been true of curriculum designers. Virtually all the Holocaust curricula designed and adopted in America violate at least one aspect of the uniqueness claim. Some have considered this an unavoidable consequence of the "Americanization" of the Holocaust.[49] While this may be true, designing Holocaust memorials should not be confused with designing Holocaust curricula. The teachers who designed these curricula were not interested in establishing a fixed interpretation of the event meant to resonate with an entire nation. Rather they were interested in using history to achieve their education objectives with their particular students. As we have seen, the uniqueness debate did not emerge spontaneously, but instead grew out of a pedagogical debate about how to present the event to Jewish students in a meaningful way. The uniqueness claim emerged as the Holocaust shifted to the center of both the Jewish consciousness and educational agenda.

The Back-to-Basics Movement

The NBC miniseries clearly marked a turning point in American's knowledge of the event and in the proliferation of Holocaust education. First of all, the show reaffirmed the use of the term Holocaust to refer to the destruction of the European Jews, rhetorically shifting the word from a metaphorical representation to a concrete reference of a specific historical event. Newspapers no longer had to use the qualifier "Nazi" in front of the word, nor provide a definition of the term in context. After April 1978, "the Holocaust" became a term that would immediately signify the Jewish experience in World War II in the minds of most Americans. Its use in reference to something else would be inappropriate. Second, the series visually reinforced the unique situation of the European Jews under the Nazis. It introduced images of concentration camps, death camps, and ghettoes in way that made the Jewish persecution distinct from other Nazi victims.

Third, the series aired at a perfect time for those already working toward increased Holocaust consciousness. It is difficult to determine what would have happened to Holocaust educational efforts had the series not aired. The evidence suggests that the topic would have continued to rise steadily nonetheless. But the miniseries provided an immediate spark to an effort that had been underway for years. Recall that New Jersey teacher Richard Flaim had been experimenting with teaching the Holocaust successfully to public school students for years before *Holocaust* aired. But when it did, parents called the New Jersey Board of Education demanding that their children learn more about it. This demand motivated the Board to work with Flaim to develop a statewide curriculum. Similarly, Roselle Chartock had also been teaching the Holocaust to her students for years, but the miniseries helped expose her unit to a national audience. Her curriculum appeared in the April issue of *Social Education* to correspond with the show.

The popularity of the *Holocaust* miniseries did not inspire Holocaust education, but it did create a greater demand for it. The materials that were already in circulation or under development would receive a political and often monetary boost as a result of the show. These materials enabled educators to quench immediately the emerging thirst for knowledge about the event. Therefore, the miniseries and phenomenon of Holocaust education reinforced each other. However, while the topic of the Holocaust gained popularity in the curriculum and American culture in the late 1970s, another movement was taking hold that, in some ways, would undermine the new interest in teaching the event. Holocaust education would continue to flourish in the 1980s, but it would do so in a far less accommodating environment.

In 1980, Republican Ronald Reagan was elected to the U.S. presidency on a rising wave of neo-conservatism, which quickly affected the educational climate. Frustrated by the waves of unsuccessful curricular reforms of the 1960s and 1970s, critics began to voice complaints about the diminishing quality of American education, especially at the secondary level. The critiques came externally from concerned parents, conservative interest groups, and politicians who worried that the curriculum had been so watered down to meet the needs of minority interest groups and liberal activists that academic standards had fallen to a crisis level. Critiques also came internally from educational researchers, who agreed that many of the reforms of the past two decades had been hastily implemented and ineffective. The result was a backlash against the neo-progressive ideas of the so-called education experts, and a return to the so-called common sense ideas of mainstream Americans.[50]

The social studies was hit particularly hard by the more conservative political and curricular climate. Researchers bickered among themselves about the origins of the current misdirection. During this time, there was an increased interest in curriculum history as historians either tried to re-familiarize their colleagues with the original intentions of the social studies founders, or tried to demonstrate how many of the attempted reforms of the 1960s and 1970s had been unsuccessfully tried before. Historian Hazel Whitman Hertzberg concluded that the social studies "reform movements of the past two decades have been more cut off from their own past than have any other major movements in the past century."[51] This sense of confusion left the social studies open to criticism and attack.

The media ran articles on the discontented public, creating what become known as the "back-to-basics" movement. Highly publicized statistical studies demonstrated the rapid decline in achievement levels and SATs scores in high-school students. As a result, disgruntled parents, often led by conservative interest groups, demanded the abandonment of affective and inquiry-based learning and the return to teacher-centered instruction. They demanded that teachers stop concerning themselves with student attitudes and values and return to transmitting factual knowledge. The system of electives and curricular freedom in most comprehensive high schools, which were implemented to better meet the demands of a diverse student population, were now considered by many as being undemocratic. All students should receive the same education, critics demanded, regardless of class, ethnicity, or academic level. The social studies lack of direction was underscored by NCSS's decision to jump on the back-to-basics bandwagon by dedicating two issues of *Social Education* to the movement in the late

1970s and by shifting its curricular emphasis in a more traditional direction. NCSS seemed to be denouncing the very reforms they helped administer. Still, other social studies researchers considered the emerging concern over eroding academic standards unwarranted and continued to push their progressive approaches to history.[52]

The back-to-basics movement received an enormous boost with the publication of *A Nation at Risk* in 1983. The National Committee on Educational Excellence, a body appointed by Ronald Reagan's secretary of education, compiled this report. The Report asserted that educational standards had fallen so low that the country was on a brink of a major crisis from which America may never recover. "If an unfriendly foreign power has attempted to impose on America the mediocre educational performance that exists today," the Report concluded, "we might well have viewed it as an act of war."[53] The Committee suggested increasing academic rigor for both teachers and students and concentrating on a core of content basics. In response to the report, many state and local governments formed committees to investigate the educational conditions of their state or community. Often their suggested remedy was the drafting and implementation of standardized testing, a movement that would continue to gain momentum in the years to follow.[54]

In the 1980s, William Bennett, secretary of the U.S. Department of Education under President Reagan, also targeted the moral education spawned by the affective revolution. In its place he and like-minded critics suggested a return to character education, which basically meant the pre-progressive "bag of virtues," transmission approach. "Moral education without justified moral content is most likely to be perceived as a pointless game," Andrew Oldquist insisted. Instead students needed, "the actual acquisition of morality, not just chatter about morality." These conservatives suggested that most Americans adhered to core of "common sense" values about how citizens should act, and that these values should be directly taught. Bennett insisted that complicated cognitive theories were not necessary, "we don't have to invent new courses. We have a wealth of material to draw on—material that virtually all schools once taught to students for the sake of shaping character. And this is material that we can teach in our regular courses, in our English and history courses."[55] Bennett published a collection of morally laden stories in his popular *Book of Virtues.*

Recall that Kohlberg considered the direct transmission of values as inherently relativistic. Having volunteered for service in World War II, Kohlberg helped smuggle Jews from Europe to Palestine. His moral theory grew largely out of his experiences with Nazism. For this reason, Kohlberg was uncomfortable with the idea of the state deciding what specific values need to be transmitted. But, more importantly, he shared with John Dewey the idea that directly teaching values would be superficial and ineffective. Meaningful values, he explained, needed to be internalized and absorbed into existing cognitive structures. However, such a theory was too complicated for most Americans to understand. As a result cognitive developmentalism and values-clarification were easily ridiculed and dismissed.

Despite the emerging curricular conservatism of the late 1970s and early 1980s, Holocaust education continued to gain popularity. Besides the aforementioned New Jersey curriculum, numerous other states appointed their own Holocaust commissions or designed their own Holocaust guidelines and curricula. Americans across the country and across the political spectrum seemed to agree that the Holocaust was a major historical event that had been unjustly neglected. While the curricular zeitgeist of the 1970s gave birth to Holocaust education, it was becoming clear the movement was not dependent upon that context for survival. But, questions over what particular version of the Holocaust should be taught continued to plague the movement. In the 1980s, that issue would became more politicized. In the next section we return to our Holocaust educational pioneers to explore the public reception of their curricula.

Holocaust Education on the Defensive

In the wake of the Holocaust miniseries, more attention was directed toward the curricula already in use, including Rabbi Zwerin's *Gestapo* simulation board game. While such an approach may have seemed like it was trivializing the suffering of Holocaust victims, Rabbi Zwerin claimed that he did not receive any objections from Holocaust survivors, some of whom had actually participated in playing the game. Not only had they played the game, but ironically, Zwerin explained, "survivors are the poorest at surviving it." This was quite logical, the Rabbi argued, because "it was sheer dumb luck in most cases that people survived... a time-out place [i.e., a place to hide] every once in a while, a piece of bread that wasn't expected... that sort of thing that kept people alive another day."[56] The *Gestapo* game attempted to recreate the random nature of the Nazi atrocities, but also demonstrate that the circumstances of the Holocaust stretched one's personal values to the limit. This was the component that was most relevant to students' contemporary concerns.

During the 1980s, the *Gestapo* game became popular with Jewish teachers, although it was not Alternatives in Religious Education's (ARE) best-selling item. Through the distribution of Social Studies Services, the *Gestapo* game managed to reach public school teachers as well. Zwerin estimated that they sold thousands of copies. The best-documented case of the *Gestapo*-based game in a classroom appeared in an ethnography by Simone Schweber (discussed in chapter six), who meticulously described the Holocaust course of Ms. Bess (Schweber's pseudonym), a teacher in a California secondary public school who used Zwerin's game as starting point for a semester-long Holocaust enactment in which Bess served as Hitler and her students as Jews.[57] Bess's course, however, went well beyond the original intent of Zwerin's game. Nonetheless, by the mid-1970s, there was a growing perception that such Holocaust enactments were being implemented.

As early as 1975, Diane Roskies complained, "The gimmicks which have been devised for emotional commemoration are numerous: In Boston summer camp the older kids staged a pogrom against younger children." There is little documentation to confirm such activities, yet a lot of anecdotal evidence suggests that affective and experiential learning techniques like the *Gestapo* game were gaining ground in the Jewish classroom. One example of a Holocaust-related reenactment was part of the curriculum for Orange County Public Schools in Florida. The activity, entitled "After the War and the Election of 2020," had students engage in a mock election of candidates that resembled those who Germans would have voted for in 1933. The objective was to trick students into voting for Hitler, and thus answering the question, "How could anyone vote for Adolph Hitler?"[58]

In her history of Holocaust education in Jewish schools, Rona Sheramy describes the growing popularity of such activities and even made reference to Zwerin's game.

> A popular method of achieving such identification was through Holocaust reenactments and simulation exercises. Accounts of such activities appeared in Jewish educational magazines and conference reports throughout the 1970s. For example, in 1973, one school described hiring a psychiatrist to set up a confinement scenario for its students. Another school used an exercise called "Gestapo" which, it claimed, aroused student interest and involvement far more than reading texts.[59]

However, the most conspicuous reference to *Gestapo* appeared in the 1980 *New York Times* article outlining the "new debate on the Holocaust." The article reported how the popularization of the Holocaust following the NBC *Holocaust* miniseries had compromised the sacredness of the event. Holocaust reenactments were presented as examples of this phenomenon. The writer quoted Max Nadel of the American Association for Jewish Education as saying that turning "classrooms into concentration camps" was "excessive," and presented the "Gestapo" game as the archetypal example.[60] The article even implied that Zwerin's game was designed to cash in on the popularity of the event, which, as we have seen, was not the designers' original intention. Nevertheless, by the late 1970s, "Gestapo" became a catchphrase for any kind of concentration camp simulation, including those exercises like classroom enactments that expanded upon Zwerin's original design. This was likely because these exercises, like the one described by Schweber, often bore the name of Zwerin's curriculum without actually following the exact outlines of his game. It is not likely that the above authors who either criticized or referred to the *Gestapo* game ever actually saw the original curriculum. Nonetheless, references and criticisms of the *Gestapo* game and other simulation exercises would not abate in the years that followed.

The game would be used as an example of the popularization of the Holocaust that blatantly denied the metaphysical uniqueness of the event. Elie Wiesel, the *New York Times* article suggested, was the leading proponent of this view. The article reported how Wiesel felt that the proliferation of sensationalized books and popularized television programs and films has dishonored the victims and rendered the public insensitive to the tragedy. He pleaded, "We need to regain our sense of sacredness."[61] Making the Holocaust into a game not only denied the sacredness of the event, Wiesel complained, but by suggesting that students could "understand" what the Holocaust was like denied its existence as an unresolved mystery. Zwerin later defended his curriculum against such attacks. "They think games turn the Holocaust into something banal," he explained, "this particular simulation doesn't turn anything into something banal...It's the facts of the Holocaust turned into an experience that people can use to understand what the Holocaust is all about."[62] Scholars would continue to debate what it means to understand the Holocaust, but the very fact that Zwerin's game attempted to transmit any form of understanding set him apart from his particularist critics.

The *Gestapo* game was a product of its socio-curricular context. The idea of reproducing the value judgments of the Holocaust victims represented the most extreme example of the affective approach to the teaching of history. But this approach was not completely out of sync with the mainstream Jewish educational literature of the time. Like their non-Jewish counterparts, certain Jewish educators were freely making connections between the Holocaust and contemporary non-Jewish events. These educators were also promoting a Holocaust curriculum that would speak directly to students' values and emotions. Pubic school teachers were also using simulation exercises to teach moral lessons. The most notorious example was that of a mid-western elementary school teacher who pitted her blue-eyed and brown-eyes students against each other in a role-playing exercise designed to teach about prejudice.[63]

A preoccupation with prejudice, values, and moral behavior permeated the secular educational literature of the time. Rabbi Zwerin even cited the theories of "values-clarification" as an important influence on the design of his *Gestapo* curriculum.[64] While the increased interest in the Holocaust by Jewish elites was certainly a result of the social upheavals of the 1960s and 1970s, this interest was exacerbated by a perceived crisis in the American Jewish community.

Jewish leaders worried that American Jews were over-assimilating and acculturating themselves in American life through social mobility and intermarriage. Some suggested that the "Americanization" of Jews was a silent cancer eating away at Jewish identity with the potential to be more destructive demographically than the Holocaust itself.[65] Holocaust education was presented as an antidote to this phenomenon, but educators argued over the root causes of the Jewish crisis, and this, in turn, influenced their pedagogical approach. Jewish particularists, such as Elie Wiesel, Emil Fackenheim, and Norman Broznick, considered the secularization of Jewish Americans as the major problem. They thought that emphasizing the spiritual aspects of the Holocaust and stressing the uniqueness of the Jewish narrative in history would inspire students to embrace their Jewish faith more enthusiastically.

On the other hand, certain teachers such as Rabbi Zwerin thought that this particularist approach was, in fact, the root cause of identity erosion in Jewish youth; the traditional teaching materials were boring and driving the Jewish youth away from their own culture. Zwerin argued that more contemporary materials that spoke to Jews as young Americans would be a more effective way to inspire Jewish identity. If they were interested in the material, they would be more responsive to it. The identity-promoting objectives and affective learning techniques would reinforce their value commitments. The *Gestapo* game, the Rabbi hoped, would engage students' interest and serve as a springboard for an exploration of their Jewish faith.

As we can see, assessing Holocaust curricula for their appreciation of the events' uniqueness is not a matter of "either/or" or "to what degree." A Holocaust curriculum can be quite respectful of certain elements while in gross violation of others. The *Gestapo* game preserved the definitional uniqueness of the event by focusing only on the Jewish experience; it conveyed the incremental steps of "the war against the Jews," with no mention of other Nazi victims. The game also preserved the historical uniqueness of the event. It focused solely on the Jewish persecution and did not approach the event as a case study in genocide. Although the discussion questions suggested that students come up with analogous examples of "Holocausts" in the present, the Jewish persecution was considered the paradigmatic event against which all others must be compared. The metaphysical uniqueness of the Holocaust, however, was violated by the very nature of making a game out of the event. The Holocaust was not presented as a sacred mystery, or unfathomable question, but rather as a historical occurrence that could be grasped like any other. Critics of Holocaust education, however, did view things this way.

Upon its publication in 1983, according to Richard Flaim, New Jersey's *The Holocaust and Genocide: A Search for Conscience—A Curriculum Guide* received universal positive feedback from teachers. The designers had anticipated objections from German American groups, like what had occurred in New York and Philadelphia, but, to their surprise, there were none. But they did receive criticisms from other groups, including Polish Americans, Turkish Americans, Armenian Americans, the homosexual community, and the Catholic League for Religious and Civil Rights. The Polish Americans objected to the way that the curriculum allegedly portrayed Poles as anti-Semitic and as supportive of the Jewish murders. The Catholic group protested the depiction of Pope Pius XII and the inclusion of certain articles that implied the Church's role in the Holocaust. Despite the fact that Flaim was Roman Catholic, the group denounced the curriculum as anti-Catholic. The homosexuals and Armenian Americans asked for more information on their groups in the anthology—requests that were honored in the second edition of the curriculum. But the Turkish American protests proved to be the most difficult to negotiate.[66]

The Turkish Americans objected to the use of the term "genocide" to depict the Turkish persecution of the Armenians in World War I. To this day the Turkish government had never officially admitted to genocide, preferring to call it a "civil war." Initially, the curriculum designers defended their position. "We will not rewrite history for any group," Reynolds remarked. "We put the facts out there and let the kids draw their own conclusions." Harry Furman was encouraged by the controversy. "Our point of this book more than anything else is to provoke debate."[67] But the Turkish Americans were adamant that the term be removed, and they used whatever resources they had to exert pressure. According to Flaim, the ADL was approached by someone in the Turkish American community with a "veiled threat"—if the Armenian genocide was not removed from the curriculum, the Turkish government would revoke its policy of providing "safe passage for Jews coming out of Iran."[68] Flaim and his co-designers were "incredulous" that "the three little pieces" on Armenian genocide in their anthology "could have an impact on foreign policy." In the tense weeks that followed, the authors discussed the possibility of removing the pages on the Armenians from the second edition of the curriculum, but they ultimately decided that the most responsible thing to do was to find another publisher, which they did. The second edition of the curriculum was published in 1985 with "sensitivity to some of the concerns of various individuals," but with the articles on Armenian genocide included.[69]

The designers of the New Jersey curriculum were driven by their desire to make the topic of genocide resonate with their students. When confronted with political pressures, they defended their curriculum on pedagogical grounds. The designers proudly shared their experiences of how the Holocaust had had dramatic effects on the attitudes and perspectives of their students of all races and ethnicities. Nonetheless, their focus was not on evoking emotional responses, but rather on arriving at a rational explanation for why the Holocaust occurred. For this reason they approached the Holocaust from a behaviorist perspective, as a study into the nature of man. *The Facing History* curriculum took a similar approach, but would be subjected to far more public scrutiny over the course of the 1980s and 1990s.

Recall that in 1980 the U.S. Department of Education recognized *Facing History* as an exemplary model education program and added it to its National Diffusion Network for use in schools across the nation. In 1982, the curriculum was published by the Foundation for national distribution. Up to this point, the privately run Facing History and Ourselves Foundation had experienced little controversy, because its curricular approach was voluntary. Even when it was added to the National Diffusion Network, there were few objections to the curriculum's design. However, by the early 1980s, the future direction of the Foundation was in dispute. Parsons wanted the materials to continue its focus on the historical particularities of genocide, while Stern Strom wanted to move in a more social activist direction.[70] In 1987, they parted ways when Parsons left the Foundation. He would later be hired as an educational consultant for the U.S. Holocaust Memorial Museum, where he continued to defend the curriculum. Stern Strom soon realized that the Foundation would need federal funding to keep pace with its widening ambitions. With the encouragement of the National Diffusion Network, the Foundation applied for a $70,000 grant from the Department of Education in 1986. What followed launched the curriculum into a national debate that would prove how politically charged the issue of teaching the Holocaust could be in the new conservative climate.

Since the Department of Education had recommended that Stern Strom apply, she assumed that she would receive the grant. Things would not be this easy. Before the grant was awarded, the curriculum had to be considered by the Recognition Division of the Department of Education, which had recently come under the leadership of Shirley Curry, who had formerly been the head of the Eagle Forum, a conservative activist group. Curry decided that the *Facing History* curriculum needed to be reviewed by a carefully selected "significance panel." The panel was headed by Dr. Christina Price, a political scientist at Troy State. In a shocking and bizarre conclusion, Price rejected the curriculum because it gave "no evidence of balance or objectivity." She concluded that "the Nazi point of view, however unpopular, is still a point of view and is not presented, nor is that of the Ku Klux Klan." She chided the curriculum for concentrating only on genocides in Germany and Armenia, while leaving out more recent examples such as the USSR, Afghanistan, Cambodia, and Ethiopia. "It is paradoxical and strange," she added, that "the methods used to change the thinking of students is the same that Hitler and Goebbels used to propagandize the German people." Veiled beneath these objections, there was a general critique that the curriculum was too liberal and too critical of America and its Cold War allies. Stern Strom was assured that this objection was not representative of the Department of Education, and she was encouraged to apply the next year.[71]

In spring of 1987, *Facing History* applied for the grant and again it was rejected. This time panelists charged the curriculum with "psychological manipulation, induced behavior change and privacy-invading treatment." They considered it "profoundly offensive to fundamentalists and evangelicals, [designed] to induce a guilt trip" and too dependent upon "leftist authorities." The next year, Stern Strom applied a third time, only to discover that the history category of the grant had been completely eliminated after her proposal had been submitted. In response to this, Representative Ted Weiss of Manhattan wrote a letter to the secretary of education calling the decision to deny funds to *Facing History* anti-Semitic. Stern Strom publicly defended her curriculum, commenting, "What in the world is the view of the Nazis, that it's good to murder people. Facing History does not teach that that's an appropriate point of view."[72] While the rejection of the curriculum certainly owed much to the political influence of the New Right, *Facing History* may have also been too progressive for more mainstream Americans as well. Nonetheless, upon its fourth review, *Facing History* was awarded a renewable $59,000 grant in 1989.[73]

In many ways, the curriculum was a relic of the curricular zeitgeist of the 1970s. The social and educational climate had moved away from in-depth investigations of specific topics, interpretive skills, and value conflicts. Many parents wanted their students to learn the "basics." By the 1980s, the Holocaust was considered an important enough topic to be considered a "basic," but it was a matter of what kind of meta-narrative one wanted to place it in. Liberals considered the Holocaust an example of an ecumenical prejudice and a societal indifference that still permeated American society. Conservatives considered the Holocaust an example of what could occur under a totalitarian regime such as the USSR. Very few argued that the Holocaust should not be taught; it was simply a matter of how it should be framed.

To a large extent the *Facing History* approach, like the other curricula mentioned earlier, reflected the basic precepts of the social studies outlined during the progressive era; its content was arranged around student interest, its historical examples were organized thematically around genocide and human behavior, and students were encouraged to participate and act upon current issues. The *Facing History* curriculum was designed in accordance with the latest educational research, much of it originating from Harvard University. In this sense, the objections to *Facing History* were actually objections to the social studies approach to history. This point was not missed by contemporaries, including historian Lucy Dawidowicz, who wrote a scathing critique of Holocaust education in *Commentary* in 1989, with the *Facing History* curriculum as one of her main targets.

Dawidowicz was one of the leading Holocaust scholars in America. Her books *War against the Jews* (1975) and *The Holocaust and the Historians* (1981) were classics in the field, and her work was quoted in the introduction to the *Facing History* curriculum. In the midst of the grant controversy, Dawidowicz was approached by an editor of an educational newsletter to write a statement of support for the curriculum. However, after reading it, she affirmed the Department of Education's decision, calling the curriculum "a vehicle for instructing thirteen-year-olds in civil disobedience and indoctrinating them with propaganda for nuclear disarmament," and considered the curriculum a representative example of how history had been replaced by the social studies. "History itself," she complained, "is under general beleaguerment in the secondary schools . . . most schools now offer a subject called 'global studies,' . . . an omnium-gatherum of pop anthropology, sociology, geography, history, and art appreciation." For Dawidowicz, *Facing History* represented the kind of watered-down, liberalized "propaganda" that had replaced "real" history in the schools. In the article, she criticized *Facing History,* along with several other curricula, for attributing the Holocaust to a more general concept of prejudice—"a generic term for hostile prejudgments of people and groups"—instead of anti-Semitism. The major problem with approaching the Holocaust from a behavioral perspective, Dawidowicz pointed out, was that it overlooked or deemphasized the historical particularities that actually caused the event to occur; a specific form of historical anti-Semitism, not a generic prejudice, was a necessary precondition for the Holocaust.[74]

Dawidowicz also questioned the benefit of comparing the totalitarian context of the Holocaust to the democratic context surrounding contemporary issues such as Vietnam, Watergate, and nuclear proliferation; they were hardly analogous. "As anyone knows who has studied totalitarian societies," she wrote, "the critical ingredient of these societies is not obedience, but terror." The actual lesson of the Holocaust, she argued, was a lot simpler than teachers had proposed. It was the Sixth Commandment, "Thou Shalt Not Murder." For *Facing History,* Dawidowicz would not be the only historian to attack their curriculum.[75] A congressional incident in 1995 would rekindle this controversy, provoking the criticisms of others.

Controversy Rekindled

In 1995, House Speaker Newt Gingrich hired Christina Jeffrey, an old colleague at Kennesaw State College in Georgia, as house historian. When it was discovered that eight years earlier Jeffrey (when her name was Christina Price) had refused a federal grant to *Facing History* for not including "the Nazi point of view," Gingrich instantly fired her.[76] The incident stirred up another media storm around the *Facing History* curriculum, and more commentators weighed in on the objectives and effectiveness of the course. The most notable was Holocaust historian Deborah Lipstadt. In an article for *The New Republic* called "Not Facing History," Lipstadt reiterated many of the objections made by Dawidowicz six years earlier.

Lipstadt blasted the curriculum for suggesting that the Holocaust has the same basic principles as nuclear proliferation, as well as attributing the event to an ecumenical prejudice. "While positing the Holocaust as unique," she complained, "Facing History presents mass murders in Cambodia, Laos, Tibet and Rwanda as examples of the same phenomenon...and adds points of comparison such as the My Lai Massacre." She suggested that while the authors of the curriculum wanted their course to be used by a wide audience "they fell prey to the same tendency they are anxious to eradicate: rationalizing evil." Like Dawidowicz, she stressed that the specific form of German anti-Semitism was a crucial part of the story, without which the Holocaust could not have occurred. To subsume that ideology in a general prejudice, "as just one in a long string of inhumanities," was historically inaccurate because it sent the message that "every ethnic slur has its roots in the seeds of a Holocaust." Along with Dawidowicz, Lipstadt seemed to be promoting a traditional approach to the Holocaust—just relate the historical particularities of the event, and let any comparisons or lessons remain implicit.[77]

Similar attacks broke out in newspapers across the country. Most writers applauded Gingrich's firing of Christina Jeffrey, whose comments on not presenting the Nazi point of view were considered absurd. Jeffrey later wrote an editorial for the *Washington Post* to "correct the record." Although she still believed that the curriculum had a "left-wing political bias," she defended herself against accusations of anti-Semitism and asserted that she was not against Holocaust education, just "the educational pedagogy" that Facing History employed.[78] Despite their disgust with Jeffrey, many writers also attacked *Facing History* for its progressive approach to the Holocaust. While the curriculum certainly included affective elements, its primary approach was progressive. Nonetheless, a reviewer for the *Washington Post* considered the program "manipulative, if one means that it seeks to affect emotions and values rather than merely imparting knowledge." In fact, a 1982 comparative research study found *Facing History* the least emotive of the four Holocaust curricula it investigated.[79] Regardless, the critic concluded that since the curriculum contained a "strong emotional component... with high-power, affect-laden narratives, films and first person accounts," students were "reshaped in the process."[80]

A writer for the *Boston Globe* suggested that *Facing History* pushed "students to see contemporary America as a latter-day Weimar Republic, slipping down the slope that leads to Dachau," and it encouraged students to think "there's a little Nazi in all of us."[81] This charge may not have been too far off the mark. The same 1982 study recorded one African American student who, after learning about the Holocaust, proudly concluded, "I guess the bottom line is there's an Adolph in me and an Adolph in you."[82] These critiques suggested that rather than helping students to think critically about history, the curriculum was indoctrinating students with a liberal agenda—teaching them to think critically about the United States. The designers of the curriculum would have proudly agreed with this charge of making students think critically about current issues. Teaching history for transformation instead of for mere transmission had been central to the social studies for decades.

The transformative power of the Holocaust was a point on which all teachers could agree. But the idea of using history to "liberalize" students made curricular and political conservatives as well as many historians uneasy. The *Facing History* controversy provided a concrete example of the century-long theoretical discussion about the most effective way to teach history. Savvy commentators such as Dawidowicz saw the curriculum for what it was—a representative example of the progressive, social studies approach to teaching history. Therefore, she not only attacked the curriculum, but also the underlying assumptions that engendered it. But others considered the curriculum some sort of excessively liberal aberration. They attacked it for inspiring civil disobedience and instilling irrational fears of nuclear war and contemporary American Holocausts. But no one criticized the fact that American students needed to learn about the Holocaust; on this basic fact, all could agree. The next chapter further explores this paradox. Although Holocaust education continued to grow in popularity based upon a consensus that the subject was important for students to learn, exactly how to frame and teach the event continued to be source of contention.

Chapter 5

Holocaustomania

The years between the two great Holocaust-related events—NBC's *Holocaust* and *Schindler's List*—were a time of steady growth for Holocaust education. Many states followed the state of New Jersey's lead in forming Holocaust commissions. The popular media covered Holocaust-related events with greater interest—although, for better or for worse, journalists tended to focus specifically on the seemingly bottomless pit of controversies surrounding the event. At the level of higher education, intellectual interest in the Holocaust became more specialized during these years. The first academic chair in Holocaust history was established at Yeshiva University in 1976. Another chair was established at University of California at Los Angles (UCLA) in 1979. Numerous more were established in the 1980s and 1990s. This was in addition to the numerous of departments of Judaic or Hebrew studies around the country, which often focused more on their interests on the event. "Yes, absolutely the chairs have made the Holocaust a special domain," Saul Friedlander, chair of Holocaust studies at UCLA, explained, "but there is no choice because otherwise it is not taught in any significant way."[1] These Holocaust experts provided journalists with juicy, controversial quotations to fill their pages. As we shall see, in general, at any given time there were those who felt the event was being neglected in American schools and needed more attention. In contrast, there were those who worried that the rush to educate all about the Holocaust was leading to watered-down curricula and Holocaust fatigue.

The Holocaust in the Popular Media

In the years following the Holocaust miniseries, the mainstream media became more interested in the Holocaust. But writers were not as fascinated with the event itself as they were with the controversy surrounding its uniqueness. In 1979–80 the *New York Times*, *Newsweek*, and *Washington Post* each ran an article on the rising popularity of the Holocaust, and how certain Jewish leaders were growing uneasy and offended by what *Newsweek* called "Holocaustomania."[2] While scholars had been arguing about the exact relationship between the Holocaust and Jewish culture for decades, particularly in regards to how to teach the event, these articles gave the impression that the "uniqueness" controversy had just recently erupted in response to the *Holocaust* miniseries.[3] Although the miniseries provided a common reference point that linked these debates to non-Jewish readers, many of the examples the articles presented to exemplify the "new" phenomenon actually predated the show. For example, the Holocaust centers that had recently opened, such as the ADL's Holocaust Research Center and the Simon Wisenthal Center for Holocaust Study, had both been established before the miniseries aired, when few Americans even knew what the Holocaust was.[4] In addition, both *Newsweek* and the *New York Times* cited Zwerin's *Gestapo,* which had also been designed before the show aired, as an example of how popular and trivial the event had become in schools. The *New York Times* quoted one unidentified teacher as saying, "If NBC could do it, if they could create fake gas chambers for their audience, why can't we do it?" One could not deny a rise in Holocaust awareness in the years following the miniseries, but the institutional and theoretical interests in the event had a long, steady history. The new issues of Holocaust uniqueness had been disputed for decades. Still, the newspapers provided a public forum for Jewish leaders to express their views to a wider audience.

Certain Jewish leaders feared that the Holocaust had become a surrogate for Jewish cultural history and complained that Holocaust museums and centers were replacing synagogues. Others were weary and skeptical of the Gentile interest in the event. In a 1981 article for *Commentary* entitled "Deformations of the Holocaust," Robert Alter, a professor of Hebrew Literature at University of California-Berkeley, complained how America was "buzzing with ideas for other barbed-wired extravaganzas," and he objected to how the Holocaust was being "commercialized, politicized, theologized, or academicized—all of which processes seem to occurring today in varying degree and manner."[5]

Education at both the higher and secondary levels was considered part of this process. The articles mentioned earlier included the rise in Holocaust education as part of the greater Holocaustomania phenomenon that had taken place since the NBC miniseries had aired—a perception that continued to plague the teachers who had introduced the topic years before. Scholars were certainly becoming more interested in the event, and the proliferation of Holocaust courses and centers were adding to the notion that the event transcended mere history—that the Holocaust was somehow its own discipline. It is difficult to calculate how many college courses on the Holocaust were actually being taught in the early 1980s; estimates ranged from ninety-three to seven hundred.[6] Nonetheless, there was a perception that some Americans were suddenly obsessed with the event. Holocaust education in American public schools continued to rise as well, an interest that was accompanied by a growing discourse on the pedagogical potentials of the teaching of the event.

The leading authority on the pedagogy of the Holocaust in the early years was historian and Auschwitz survivor Henry Friedlander. Friedlander moved to the United States in 1947 to earn his PhD in modern European history. In 1973, as professor of Judaic studies at Brooklyn College, he published a critique of the treatment of the Holocaust in American textbooks.[7] He was quoted generously in a 1979 *New York Times* article about Holocaust education, one of the first attempts to consider seriously the pedagogical elements of the event.[8] In an article published earlier that year in *Teacher's College Record*, a leading educational journal, Friedlander outlined his views more fully. To this day, this article is one of the most highly regarded and widely cited justifications for Holocaust education in America.

Friedlander feared that the new "proliferation and popularization" of the Holocaust could be more detrimental than beneficial to American education because certain teachers were being forced to teach subject they knew little about. He considered it "ludicrous" that large school system like New York would mandate Holocaust education without first providing adequate training to it teachers. Small systems, however, like Brookline, Massachusetts, home of *Facing History*, could be more successful. He also denounced the NBC miniseries, which he considered "poor history and bad drama" and worried that such "kitsch" could be construed as education. "There is a difference between education and mass culture," he explained, "it is an error to confuse the two." Education for Friedlander was not merely exposure or transmission of factual knowledge. It was an endeavor to understand the past as fully as possible, to meditate upon and analyze the different components of the event in an attempt to understand and explain it. Conspicuously absent from his writing were any references to the incomprehensibility of the Holocaust, or the inability of a non-witness to fully understand the event. In fact, he insisted on its comprehensibility—a view that distinguished him from many of the other survivor–scholars.[9]

He understood that in order to reach its pedagogical potential, the event would have to be universalized, and that any claims of the epistemological uniqueness of the Holocaust were antithetical to this goal. To suggest otherwise would be hypocritical. He explained:

> One cannot treat the Holocaust as sacred history and also insist that it became a lesson and warning for public discussion as well as an integrated part of our school curriculum. And throughout much of the debate about the Holocaust there is this attempt to have it both ways; to have it unique, and yet to have it as only the last example of two thousand years of persecution; to teach it as a moral lesson, and yet to make it so particular that no one else can use it. These are contradictions that must be resolved.[10]

In addition he argued that Holocaust comparisons were entirely appropriate if they were used to "understand and learn from the Holocaust, not trivialize it." The historical uniqueness of the Holocaust could not be maintained because events like the Armenian genocide, treatment of the American Indians by the United States, the transportation of black slaves across the Atlantic, the internment of Japanese Americans, the "barbarism of the Vietnam war," and the Stalinist camp systems all shared certain aspects with the Holocaust. The Holocaust was unique in bringing all these components together, but making comparisons, he suggested, would underscore this point, not undermine it.[11]

Friedlander suggested that the Holocaust was a major public event, equal to the Fall of Rome or the French Revolution. Every student should learn about it because it helped explain the conditions of the present. The Holocaust complicated modern notions about the progress of man and the inherent benefits of technological advance. It also provided a window into the nature of society, allowing "us to glimpse human behavior in extreme situations." Friedlander suggested that teachers and scholars should employ social sciences, such as psychology, political science, and sociology, in trying to explain the different dimensions of the event—as a way to gather useful data and apply it to contemporary problems. Finally, as the teachers discussed in the previous chapters had done, he linked Holocaust education with the larger goal of the social studies, teaching civic virtue. "We need to teach the importance of responsible citizenship and mature iconoclasm," Friedlander wrote, "We must show how the defense against persecution and extermination is citizens prepared to oppose the power of the state." The Holocaust could teach students about their responsibility to be politically active and aware, he argued, by pointing out the potential consequences of failing to do so.[12]

The views of Henry Friedlander demonstrated that the controversy over Holocaust uniqueness could not be cast along cultural lines. Not all Jewish historians defended the phenomenological uniqueness of the Holocaust, nor did all concentration camp survivors assert its metaphysical uniqueness.

There was a range of views within each group. Perhaps it seems counterintuitive that Friedlander, as a historian and Holocaust survivor, would be one of the strongest proponents of a "universalized" Holocaust curriculum. But, since he believed that the event could be used to foster civic virtue in American youth, he was less cynical than many of his peers about the methods and intentions of teachers. Friedlander's 1979 article provided a theoretical morale boost for the entire Holocaust education movement and helped the movement survive the Holocaustomania backlash. Indeed the movement was thriving.

The Holocaust in Textbooks

Since there was a dearth of educational material available on the Holocaust in the mid-1970s, teachers like Richard Flaim and Edwin Reynolds of New Jersey either had to use what was already in existence, such as Chartock's *Society on Trial* curriculum, or they had to develop their own. Reynolds explained, "textbooks didn't like to deal with the topic, and there was reluctance on the part of publishers to give us material."[13] These early teachers of the Holocaust would not only seek to research and design their own curricula and courses, but they would also work to disseminate their ideas and materials to other teachers. Often their initial concern was simply to provide teachers with a basic knowledge of the event because textbooks failed to do so.

By the late 1970s, some history textbooks were outright ignoring the Holocaust in their narratives, while others were not providing the particular details that made the Holocaust such an unprecedented event. They covered the event briefly in the context of World War II. A representative example, taken from a 1979 Macmillan secondary text called *A Strong and Free Nation*, read: "Few of them [Americans] knew all that had been going on in Germany under Hitler's rule. The dictator had a violent hatred for the people of Jewish faith. In Germany and elsewhere, these people were taken from their homes and placed in special prisons known as 'concentration camps.' There several million Jews died or were put to death in gas chambers."[14] This depiction ignored several important aspects of the event. It failed to mention the percentage of the total European Jewish community that was murdered, and the fact that many survivors were still alive and living in the United States. This account also failed to outline fully the centrality of anti-Semitism (racial, not religious) to Hitler's political and social agenda, the origins of Hitler's theories on racial hygiene, the enormity of the German's bureaucratic coordination in his attack on the Jews, and the Nazi's use of mobile killing units that shot thousands of Jews before the death camps were even created. The text was ambiguous about the knowledge of the Holocaust by U.S. authorities and Allied nations, and it avoided any mention of the anti-Jewish U.S. immigration policies. It also omitted any mention of the knowledge and cooperation of contemporary German civilians.

The title of this particular text suggests that it had a celebratory inclination toward the United States, which may have explained its failure to mention certain facts that may been construed as critical of U.S. policy or its West German ally. Nonetheless, at the time, this type of superficial coverage was common in most secondary and college texts. A study of 1970s history textbooks by Glen Pate demonstrated that the median coverage of the Holocaust was twenty lines per text. Pate concluded that textbooks "do not give the Holocaust the treatment it deserves in its own right." He also criticized the texts for failing to offer any lessons to be learned from the Holocaust and for refusing to connect the Holocaust to contemporary events. Many textbook companies would have considered such inclusions as going beyond their mandate. Textbook authors often aimed to present the facts in the most inoffensive, straightforward, objective way they could. To understand fully why textbooks might gloss over the history of the Holocaust, we must consider the nature of textbook adoption in the 1970s and 1980s.

Criticisms of textbooks were common throughout the twentieth century.[15] Attacks originated from both the Right and the Left. Conservatives considered certain texts too critical of American democracy, traditional American heroes, and Christian ideals. Liberals complained that texts did not include enough material on the contributions of minorities and problems of democracy. As a result, textbook companies often took a safe middle position that satisfied no group fully, but avoided offending any major groups. This process impacted materials for Jewish schools well. By the 1980s, the Reform and Conservative movements turned away from Jewish educational publishers toward outside, commercial publishing houses. As a result the material in these books became more ideologically neutral and homogenized.[16]

By the 1960s and 1970s criticisms of textbooks came to a crescendo. In hopes of deterring the future production of such dull, inadequate texts, educational researchers began paying more attention to the textbook design and adoption process. Educational theorists like Michael Apple criticized textbook policies for considering profit above their responsibility to contribute to a democratic education. Others drew attention to textbook companies as "gatekeepers of ideas and knowledge."[17] As theorists considered the relationship between knowledge and power, they worried that textbook companies had been a far more influential force in controlling the curriculum than previously thought. With estimates that 75 percent of classroom time and 90 percent of homework time was spent engaged with a textbook, researchers became critical of textbook content and adoption procedures.[18]

The perception that textbook companies were acting irresponsibly was reinforced by the fact that in the 1970s independent publishing companies were rapidly being bought out by larger conglomerate firms such as IBM, Xerox, and RCA. The larger firms aimed their products at national markets, and so their books had to appeal to all regions of the country. But, in many cases, the content of their books were often guided by the textbook adoption policies of a few specific states. Some states, such as New York and Massachusetts, had local selection boards, where school districts could choose their own materials for use in the classroom. In other states, such as Florida and Virginia, school districts had to choose from a list of texts approved by the state. But in certain states such as Texas and California (for grades K–eight), local schools had little control over their textbook selection because the state adoption committee selected specific textbooks that would be used exclusively by the entire state. Local districts often had to decide whether to adopt the required texts, or forgo state funding. Therefore, to capitalize on the enormous potential market in Texas and California, textbook companies often catered their content to fit the needs of these two constituencies.[19] As a result, critics argued that interest groups from these two states had a disproportionate amount of influence on the content found in texts.

Despite the research and political maneuverings of these textbook companies, the enormous amount of money they invested in designing and marketing their products made the industry quite risky. They had to find the proper balance between the Right and the Left. The Civil Rights movement not only inspired blacks to fight for their own equality in the curriculum, but it also encouraged other minority groups, including the Jews, to pursue their own representations in the curriculum. These minority interest groups often mobilized their resources both regionally and nationally to attack textbook companies. But in response to minority gains in the 1960s and 1970s, conservative parents across the country demanded a return to the basics. They not only criticized the new multicultural material, but also many of new affective learning techniques that accompanied it. "Educators no longer worry about whether a child can write," one parent complained. "They worry about what the child writes, what the attitude is towards a particular subject."[20] Textbook companies were left to balance these two extremes. In 1975, the *New York Times* commented how "There are now two sets of pressure groups—the old one on the Right and the new one on the Left. Textbook publishers are struggling to catch up with the second without leaving the first too far behind."[21] The consolidation of the textbook industry during the 1970s left these companies with more financial resources but less flexibility in responding to regional and local concerns. So did this prevent the Holocaust from receiving adequate coverage in textbooks?

The answer is both yes and no. In the 1970s, the Holocaust was only being taught in the Northeast, in states with local school adoption committees. Therefore, it made no financial sense for textbooks to provide anything but a superficial coverage of the event, especially if texts now needed to devote more of their space to other minority groups with more vigilant lobbying groups. The Jewish community had consistently been critical of textbook treatment of the Holocaust throughout this period, but their objections were often inconsequential. As early as 1961, in an article for the Jewish periodical *Congress-Bi-Weekly,* Gerald Krefetz criticized the coverage of the Holocaust in American textbooks. That same year the New York City Board of Education issued a letter to publishers concerning "serious deficiencies in textbook treatments of Nazi atrocities against minorities."[22] This was followed by similar surveys and critiques by the Anti-Defamation League (ADL) and American Jewish Congress throughout the 1960s and 1970s. As the continued lack of Holocaust coverage in textbooks proved, these criticisms had little to no impact.

To a large extent, textbooks companies had relatively little pressure to increase their coverage of Jewish history. Jewish leaders took a more restrained approach to getting their history included in the curriculum when compared to black interest groups, who were far more aggressive and, thereby, effective at changing textbook content. For example, in Detroit in 1962, the NAACP successfully blocked the adoption of a history textbook that did not depict black Americans with dignity. After persistent protest the textbook publisher issued a newly revised edition omitting the objectionable material and incorporating new content on black history. Similarly, in 1974, the authors of a Mississippi textbook that included generous amounts of black history sued the State Purchasing Board for rejecting their book. After a five-year court battle, they won the case and their book was approved for use in Mississippi classrooms.[23] The Jewish community attacked textbook companied themselves, not the state or local adoption committees that helped perpetuate their blandness. They issued guidelines and published critiques, but mostly for Jewish audiences. As a result, they failed to muster any public support or to attract any major media attention to the lack of Jewish history in textbooks.

While the topic of the Holocaust made slight gains in the curriculum during these years, it was not a direct result of more attention to minority history. In fact, most educators at the time did not necessarily consider Jews an American minority. Educational articles on multiculturalism that appeared in the late 1960s and 1970s often failed to mention Jews at all. For example, a 1969 issue of *Social Education*, devoted to the study of minorities, included "American Indians, the US Hispano and Orientals," but not Jews.[24] The position of Jewish Americans in the upper socioeconomic levels of American society exempted them from any claims of institutionalized economic and political oppression that plagued other minority groups. So when teaching the Holocaust gained popularity, it did so as an indirect consequence of teachers paying more attention to the lives of oppressed American ethnicities. Many teachers did not introduce the Holocaust to learn about the oppression of Jews in modern America, but instead as a way to consider the oppression of other minorities in American society.

Therefore, Jewish history did not experience any immediate gains as a result of the multicultural movement and resulting diversification of the curriculum. Instead, textbook companies catered to the more historically oppressed American minorities and their proponents. An explosion in scholarly research on those topics supported the increased attention to minority history. Jewish history simply got lost in the abundance of new material.

But if Texas and California actually exerted that much influence on the content of textbooks, as many contemporaries argued, shouldn't the Holocaust have gained more space in textbooks, since both Texas and California have considerable Jewish populations? In fact, in the 1990s, California would become one of only a few states to mandate human rights education in its schools. The more compelling argument for why textbooks didn't cover the Holocaust in detail was simply that most Americans had little knowledge or interest in the event prior to the NBC miniseries. The meager attention allotted to the event in textbooks reflected the general interest of the American public at the time. Another contributing factor was the time lag between historiographical developments and their appearance in textbooks. One scholar estimated that it generally took ten years for the textbooks to catch up with the knowledge of the scholarly field.[25] The field of the Holocaust had developed considerably since Raul Hilberg's *Destruction of the European Jews* first appeared in 1961, but it would be decades before textbooks reflected this.

Holocaust Historiography

In 1986, Hilberg reviewed the literature on the Holocaust and concluded the historiography "is sufficiently voluminous to qualify unmistakably as a separate undertaking."[26] Besides Hilberg's seminal work, Lucy Dawidowicz's *The War against the Jews,* published in 1976, became a standard text. In this work, she asserted the definitional uniqueness of the Jewish experience in the Holocaust by concentrating on Hitler's anti-Semitism. She argued that the Nazis waged World War II to allow themselves to implement "the Final Solution," as the Nazi called it. Yehuda Bauer published another important study, *A History of the Holocaust* in 1982, in which he contextualized the Holocaust in a larger narrative of anti-Semitism. These studies stressed how the Jewish question directly impacted the policies and decisions of Hitler and his Nazi leaders, and how Hitler was intent on murdering the Jews from the day he took power. This view would eventually be dubbed the "intentionalist" school of interpretation because it underscored the importance of Hitler's anti-Semitism as the driving force behind the Holocaust, even though there were no sources that directly linked the destruction of the Jews to his orders. The competing school of interpretation was called the "functionalists." These historians de-emphasized the direct role of Hitler, instead focusing on the complicated web of the Nazi bureaucracy. They emphasized the incremental nature and shifting policies of the Jewish question. The leading proponent of the "functionalist" view was West German historian Martin Broszat, who published *Der Staat Hitlers (The Hitler State)* in 1969.[27]

Initially West German historians failed to address the Final Solution directly in the years following Word War II. Instead they concentrated on explaining the rise of the Nazi state. The dominant paradigm in the 1950s and 1960s was the *Sonderweg* thesis, which connoted Germany's "divergence from the west." Proponents of the *Sonderweg* thesis examined the rise of National Socialism (Nazis) from a "structural" social scientific perspective by comparing Germany's development to that of other modern Western nations (France and Britain). Since Germany, they argued, never had a bourgeois revolution that replaced the traditional, feudal ruling elites, the German state experienced a belated modernization process, retaining many of its premodern characteristics. This inherent weakness undermined the success of the Weimar Republic in the interwar years, making Germany particularly susceptible to the rise of fascism. This interpretation helped to reconcile the Germans with the western postwar democracies and helped to paint the Nazi dictatorship as an abnormality in Western history. It also shifted focus from the decisions of individual Germans to the underlying structures of the German state. For these reasons, the *Sonderweg* thesis helped the Germans deal with their dark past. However, the thesis would come under attack in 1960s, when West Germans historians like Broszat, began to explore the Germans' role in the Holocaust more directly.[28]

In fact, it was the airing of the NBC *Holocaust* miniseries in West Germany in 1979 that first inspired its citizens to deal with the dark German past. Not only did West Germans fail to address the Holocaust, but until the 1960s they did not even study National Socialism, because many former Nazi supporters were still serving in government positions. By the late 1980s and 1990s, local governments began suggesting that its vocational and secondary teachers address the extermination of the Jews explicitly by visiting local memorials or Holocaust-related sites. They also began to list "Persecution of Jews" and "the SS and concentration camps" explicitly in their curriculum guides.[29]

It is important to note that these German, Israeli, and American historians used social scientific techniques to understand the Holocaust. One of the primary tools of the social scientist and the historian throughout the century has been the use of comparison. This technique, when applied to the Holocaust, sparked reactions from historians like Dawidowicz and Bauer who defended the historical uniqueness of the event. In his discussion of the controversy over the memory of the Holocaust in West Germany, Charles Maier reflected, "Any genuine comparative exercise emphasizes uniqueness as much as similarity; . . . A historian's unwillingness to acknowledge both aspects of the venture and a tendency to minimize distinctions at the cost of what is similar may indicate a partisan intent."[30] Comparison, according to Maier, is a necessary tool in discovering the particularities of a historic event.

Any restriction on historical comparisons threatened the open-minded inquiry underlying the entire process of knowledge construction. So comparing the Holocaust with other events or comparing the Nazis with other totalitarian governments was not an idea introduced solely by social studies teachers as a means of making the event more relevant to their students. It was also a tool used by leading historians and social scientists in the field. Of course, comparing historical events could potentially "normalize" them by de-emphasizing the particularities of a specific event to the point where it appears like a natural impulse or logical progression and, thereby, taking away human agency. Comparisons could diminish the guilt of the perpetrators by making their actions seem common. Nonetheless, the problems of historical comparison were inherent in the process of writing and understanding history itself, not in studying the Holocaust as a specific topic. But, as the 1977 *New York Times* controversy proved, the Holocaust did bring these methodological and epistemological issues to the attention of the mainstream public in a way that few other historical topics had done. The first teachers of the Holocaust were aware of this debate, and felt obliged to take a position on the different elements of Holocaust uniqueness.

New Jersey's Richard Flaim explained how he thought that the Holocaust "was both unique and must be taught for its universal aspects." If the Holocaust was taught well, he explained, "and they [students] also studied other events that involve genocide, and they studied them well, the students will see both the uniqueness of each of those events as well as those things they may have in common."[31] When Flaim and his colleagues designed their curriculum in the mid-1970s, they found themselves at a point where textbooks were barely covering the topic, but the historiography on the Holocaust was developing rapidly. Researchers from West Germany and America were beginning to challenge some of the interpretations of the Israeli and American Jewish historians. Scholars were also producing more nuanced depictions of the perpetrators, victims, and bystanders and exploring the various elements of the uniqueness claim.

To overcome the gap in knowledge between the textbooks and field, the early Holocaust curricula, such as Chartock's *Society on Trial*, New Jersey's *The Holocaust and Genocide: A Search for Conscience*, and New York City's *The Holocaust: A Case Study in Genocide*, were often published with accompanying Holocaust anthologies. These resources were hundreds of pages in length and included excerpts from dozens of primary and secondary sources. They were designed to replace the textbook, not to supplement it. The increased availability of these curricula in the years following the NBC miniseries further decreased the need for textbooks to cover the Holocaust at great length. These curricula also reinforced the idea that Holocaust was somehow too important a topic to be left to the textbooks. The enormity of the event justified it own materials.[32] In addition, subsequent Holocaust curricula were expected to reflect the emerging historical research.

Mandating Holocaust Education

A 1987 national history survey found that although only 32 percent of seventeen-year-olds could place the Civil War in the correct half-century, 76 percent could identify the term "Holocaust" with reference to the Nazi genocide during World War II.[33] By 1988, eight states, New York, California, Pennsylvania, New Jersey, Ohio, Tennessee, Virginia, and Connecticut, had each developed some kind of state-endorsed Holocaust, genocide, or human rights curriculum.[34] Over the course of the 1980s and 1990s, state legislatures took a number of different approaches to demonstrate their support of teaching the event. By 2007, Illinois, New Jersey, and Florida directly mandated the teaching of the event in their public schools. California, New York, and Massachusetts embedded the Holocaust in the broader spectrum of human rights and genocide education, and required that teachers address a list of group atrocities in their classrooms in some manner. Connecticut, Indiana, Ohio, Pennsylvania, and Washington "encouraged" or "recommended" teaching the Holocaust and created commissions to support it. Georgia, Alabama, Maryland, Nevada, North Carolina, South Carolina, Tennessee, Rhode Island, and West Virginia also appointed Holocaust and genocide and/or human rights commissions to develop resources and support teachers, but did not in any way push the topic onto teachers.[35]

Of course, in most cases these mandates were not funded beyond the formation of a committee or commission, nor did states provide any kind of enforcement; they were mainly political statements.[36] Nevertheless, as we shall see in chapter seven, these commissions could make an immediate and substantial impact at the local level, but mostly with teachers who were already teaching the event. Efforts to mandate Holocaust and genocide education continued into the new century. In most cases proponents of Holocaust education opposed these efforts. They worried that most teachers will not be prepared to teach the event adequately. Pioneers of Holocaust education preferred slower more meaningful reform through workshops and the distribution of a well-designed, state-endorsed curriculum. In other words, Holocaust educators, while indeed proselytistic, preferred voluntary, not forced conversions.

The process of mandating the teaching of Holocaust steered pedagogy directly into a collision course with politics, in which cases the latter always triumphed. In 1995, the state assembly of New Jersey approved 77–80 a bill mandating the teaching of the Holocaust in its public schools. In addition the legislation named the topics of the annihilation of Native Americans, slavery, and the Armenian and Cambodian genocides. The bill was delayed by disputes over what counted as genocide and which groups to include. New Jersey teachers defended the universality of Holocaust education, suggesting that other groups did not need to be added explicitly. "[The Holocaust] usually leads to further discussion of what's going on in other places and in America with regards to racism," one teacher explained, "The students see how it applies to their own lives ... it's relevant for today when you look at what's going on in the world and how history repeats itself."[37] Nevertheless, political interest groups continued to lobby for greater inclusion.

By 1996, a full-fledged controversy had erupted in New Jersey over whether or not to include the Irish famine in the list of genocides to be covered. "We'll bring in experts from around the country and Ireland and Northern Ireland to testify about the famine to prove that this just wasn't a natural disaster, but actually a genocide," one member of the Irish Famine Curriculum Coalition exclaimed. In response several Jewish groups complained that including the Irish famine, which killed nearly a million people, would dilute the impact of the Jewish experience. "You must distinguish between that kind of tragedy and an extermination policy against a group of people that is premeditated," one Holocaust survivor pleaded, "The Irish could get on ships and emigrate, not the Jews." On the other hand, other Jewish leaders encouraged the inclusion of other tragedies because they would possibly bring more interest to the uniqueness of the Holocaust.[38]

In 1996, a similar debate erupted in the state assembly of New York, where an Irish American assemblyman from Queens sponsored a bill requiring that Irish potato famine be taught in all its public schools. "Upwards of a million people starved to death, and there was forced migration of over a million more," he explained, "to us that was the Irish Holocaust. There is no video or film to document it, but I wasn't a calamity—it was willful act of neglect that caused it." Another assemblyman, who opposed the bill, worried that such legislation would open the floodgates of ethnic histories. "We can go on and on. If we do this, we legitimize any other ethnic group demands and we would have a legislature-driven curriculum in our history classes."[39] Nevertheless, in 2001, a thousand-page curriculum on the Irish potato famine, designed by a team of educators at Hofstra University, was sent to schools across New York State. The curriculum used the famine to explore the universal themes of poverty, hunger, and starvation. Thus, New York and New Jersey, states that both began as Holocaust education pioneers, were driven down the road of universalizing the event, not for pedagogical but rather for political reasons.

However, it was a New Jersey teacher who expressed the most clear-headed thoughts on the Holocaust–genocide–human rights controversies. In 1998, in response to an editorial that blamed teachers for students' ignorance of the Holocaust, the teacher wrote the following:

> does he know that state mandates lessons in career education, affirmative action, drug education, Holocaust studies, and foreign language. Teachers are expected to teach all of the above, plus deliver a comprehensive curriculum, while children are walking in out of the room to go to speech, basic skills, physical education, resource rooms, programs for the gifted and talented, instrumental music, and computer education... There's nothing wrong with the rote work and memorization... But there is something wrong with criticizing the schools that are trying to deliver a curriculum, meet state mandates, prepare for standardized testing, and develop a love for learning and an inquiring mind in our youngsters.[40]

With increasing frequency throughout the twentieth century, the schools have been used to address virtually all of societal ills. As these mandates continued to accumulate, everything inevitably got watered down. When the political concerns of scholars and politicians met the reality of classroom, it was usually the latter that won out. As we shall see, this understanding was at the forefront of the minds of the curriculum designers in the state of Ohio.

Designing the Ohio Curriculum

The process of designing a state curriculum generally followed the pattern established by New Jersey. It involved the cooperation of a broad spectrum of interested participants such as politicians, community leaders, religious leaders, survivors, and local educators interested in disseminating knowledge about the event. But, in each case, the movement began at the grassroots level. It would be redundant to outline each state's adoption process, but through an investigation into the construction of the Ohio curriculum, the reader can get a sense of the issues at stake in formulating an official curriculum.

In 1986, Executive Order 86-24 by Governor Richard Celeste formed the Ohio Council on Holocaust Education. The Order established an official definition of the Holocaust as the murder of "six million Jews and millions of other Europeans," subtly distinguishing between the Jewish and non-Jewish victims. In addition, the Order outlined the reasons for remembering and teaching the event, also being careful not to devalue the historic atrocities against other groups. It announced, "all people should remember the horrible atrocities committed at that time and other times in history." In the final report the Council reiterated this distinction, insisting that "we must never minimize any genocide, but we must recognize that Holocaust was unique." Ultimately, of all the curricula covered in this study, the Ohio curriculum would be the most protective of the historical and definitional uniqueness of the Holocaust, but it would still refer to other atrocities.[41]

The Ohio Holocaust Council met four times in 1987 and consisted of five different Committees, including a Needs/Assessment Committee that surveyed the Ohio state schools to assess to what degree and frequency the Holocaust was already being taught. They discovered that out of the 103 schools that responded, only 4 did not teach the event at all, but only 1 school taught the Holocaust as a separate course. The vast majority of Ohio schools fell somewhere in between, offering cursory coverage of the event across different subjects and different years. The two most common explanations selected for this minimal treatment was they gave other subjects a higher priority (31/103) and that they had a minimal Jewish population (31/103). Perhaps the most significant finding was that most Ohio teachers believed the Holocaust should be universalized by including other minorities persecuted by the Nazis (36/103) and by comparing it to other tragic events in history (68/103).[42] It is difficult to determine from this data whether the Ohio teachers believed in universalizing the event for pedagogical or for cultural reasons. In other words, did they believe that studying multiple events was necessary in order to be able to make generalizations about the nature of government and man, or did they merely believe that non-Jewish students would not be interested in learning about a Jewish event unless the curriculum included other minority groups? Most likely, it was the latter. Many teachers were suspicious of a "top-down" curricular movement that emphasized one minority group above the rest. To win over these reluctant teachers, the reforms would have to be administered from the "bottom up." Accordingly, the subcommittee wisely suggested that the Council should focus on the "enrichment of existing programs," because many schools seemed to be already teaching the topic. This suggestion was heeded by the Council, which decided against mandating Holocaust education. However, it did prepare an official state curriculum entitled *The Holocaust: Prejudice Unleashed*, which was distributed to schools throughout the state in 1989 for voluntary use. Leatrice Rabinsky and Carol Danks, an odd but effective pairing of Holocaust educators, designed the curriculum.

Rabinsky is an Orthodox Jew, whose parents taught Judaic studies and whose grandfather was a rabbi. Her family had been involved in raising awareness and organizing American Jewish rescue efforts during World War II. As a child she met with Jewish refugees who had fled the Nazi persecution, and as a young woman she helped many Jewish "displaced persons" adjust to American life after the war. Danks, on the other hand, was a Southern Baptist who had little to no knowledge about the Holocaust until she accompanied her husband, a visiting psychology professor at Warsaw University, to Poland in 1978. Ironically, while Danks missed the Holocaustomania that surrounded the NBC miniseries in America, she learned more about the Holocaust that year than many of her American counterparts. Through visits to the Warsaw Ghetto Uprising memorial, Auschwitz, Birkenau, and her conversations with local residents, Danks learned about the horrific deeds that had taken place in Poland during World War II. For both Rabinsky and Danks, their first-hand experiences with those who witnessed the Holocaust, although quite different in nature, inspired them to introduce the topic to their students.

One thing the two educators did have in common was that they did not approach the Holocaust through a social studies orientation, but rather through literature. "Literary truths are as important as historical ones," Danks insisted.[43] As early as 1965, Rabinsky began integrating Holocaust literature into her seventh–ninth grade courses. This eventually evolved into a two-week unit. In 1973, Rabinsky was promoted to Cleveland Heights High School, and the next year she introduced a senior elective on the "Literature of the Holocaust." The twenty-eight students who enrolled were both Jewish and Christian. The class was so successful that at the suggestion of one of her students, she organized a "Journey of Conscience," a two-week pilgrimage to European sites of the Holocaust. The field trip was a moving and informative success, and it soon became a permanent part of the course.[44]

In 1979, upon her return from Poland, Danks organized her own course on Holocaust literature. After her colleagues had shown enthusiastic interest in the event, the class was extended to the entire ninth grade the following year. While she worried about confronting ninth-graders with such dark images, Danks reflected how her students "needed to see and know these events because they are part, albeit an ugly one, of what its means to be a part of humanity."[45] Due to their early involvement in Holocaust education in Ohio, both teachers were appointed to serve on the Governor's Council in 1987. When Rabinsky was appointed to chair the Materials and Curriculum Committee, she chose "the bright young, . . . friendly [and] cooperative" Danks to serve as coeditor of the proposed curriculum. They worked on the unit for two years before it was published and distributed in 1989.[46]

The *Prejudice Unleashed* curriculum was dedicated exclusively to the Holocaust; there was no mention of other genocides. It began its narrative with an overview of the interwar Jewish communities in Europe, so that students would not view the Jews solely as victims. The authors included lessons on the destruction of the European Jewry, ghetto life, concentration camps, labor and death camps, resistance, rescue, and postwar trials. The final section, titled "Meaning of the Holocaust in Today's World," included lessons on survival and renewal. Overall, the curriculum provided a balance between the heroic and tragic elements of the Holocaust and was careful not to put any positive spin on the event. The program, the authors hoped, would allow students to "recognize attitudes and behaviors that lead to prejudice and hate" and "understand world situations that require an individual and/ore group resistance."[47]

The Ohio case fit the pattern established by New Jersey. At the grassroots level, certain teachers, both Jew and Gentile, introduced the topic to a limited number of students. The topic was received with enthusiasm, inspiring other teachers to get involved and bring the topic to more students. At some point the state governor or Board of Education, having been pressured by teachers and parents, decided to support the growing movement by providing institutional and financial support. The governor then appointed a Holocaust Council or Committee, which joined religious, political, and communal leaders with local teachers. Collectively, they deliberated on the most effective means of expanding Holocaust education. While it may seem that generalizations are being based on only two cases, a study by Wilson Frampton confirmed these assertions.

In 1989, Frampton surveyed the Departments of Education in all fifty states to ascertain to what degree they were supporting local Holocaust education. He discovered that while only six states had developed materials for teaching the Holocaust (oddly, Ohio was not among the ones he listed), twenty-three were aware of at least one school district in their state in which the Holocaust was being taught. Of course, as Frampton pointed out, many of these states simply may not have been aware of what was and was not being taught in their constituency. The state Departments of Education that did not officially endorse Holocaust education cited that they wanted to avoid interfering with the local autonomy of school districts, and that they did not want to inspire other minority groups to lobby for the inclusion of more content. The ones that did endorse Holocaust education, on the other hand, did so in response to teacher initiative and demand. "Since most states seem reluctant to initiate special interest content proposals," Frampton concluded, "change must first be initiated at the local school district level... teachers at the local district level are the primary means for initiating a proposal to change state policy."[48] This "trickle-up approach," as Frampton called it, was the only effective way to gain the support of the state.

So why were certain teachers so interested in teaching the Holocaust, even when the state governments didn't actively support it? The answer has to do with the grassroots nature of the movement. The survey from Ohio suggested that initially teachers saw the event as a single minority group vying for more space in the curriculum, and, therefore, were resistant to adopt the topic in more than a superficial way. But once they saw the Holocaust being taught successfully by teachers in their school, and how this "Jewish event" could be used to excite and engage non-Jewish students, they were more willing to try it themselves. In other words, once they realized the pedagogical potentials of the topic and how it transcended cultural particularities, they viewed the topic in a new light. Rabinsky and Danks, as Flaim and Reynolds had done in New Jersey, first began teaching the course to a limited number of students, but the topic quickly spread throughout the school and state.

A key component of this success was the establishment of local workshops. Once teachers became aware of the universal aspects of the Holocaust in these workshops, then, in the words of Richard Flaim, "you couldn't stop them from teaching it."[49] Teaching seminars seemed to be the most important factor in initiating change. This realization didn't escape the attention of the Ohio Council. In their discussion, one teacher pointed out that "teachers will teach what they and the students want, whatever the Council recommends." Another participant noted that in Dayton, "the success and quality of Holocaust education coincided with the availability of teacher training seminars, not the number of Jewish students."[50] For these reasons the Ohio Council decided against a state mandate, choosing to support the momentum of the movement at the grassroots level instead.

Another important factor in the spread of Holocaust education was the cooperation and coordination of the various educators involved. The previous case studies demonstrated how the ADL could be effective in forming contacts among Holocaust educators like Roselle Chartock and the New Jersey teachers. Leatrice Rabinsky had also met with the New Jersey teachers early in their design process to share her experiences with teaching the Holocaust. In addition to this, Rabinsky had met with Margot Strom Stern and William Parsons at a Conference in Mercy College in Detroit in 1976, where they "exchanged materials and ideas."[51] When state curriculum writers like Rabinsky and Danks began designing their own units, they first familiarized themselves with what already existed in New Jersey, New York, and Massachusetts, and, in many cases, spoke with the authors themselves. They then chose to emulate, refine, or rewrite portions of these curricula to find an approach that best suited their state. In this sense, Holocaust education was a cumulative process, based on a network of cooperating teachers, built upon a shared body of anecdotal evidence and experience, and united by a common goal—the dissemination of Holocaust teaching materials and the training of as many teachers as possible. This network continued to expand and develop throughout the 1980s, adding to the effectiveness of the grassroots movement.

Although these ambitious teachers were the primary force behind the movement, the story would be incomplete without recognizing the role of the Jewish organizations, particularly the ADL. Most of the curricula presented in this study benefited from the organizational or financial assistance of the ADL, but the organization did not have an active agenda of recruiting Holocaust teachers. In each case, these teachers approached the ADL after they had already designed and implemented their unit. Also, the ADL consistently endorsed a universalized version of the Holocaust, which emphasized the ecumenical nature of prejudice, not the historical particularities of anti-Semitism that would have made the Holocaust a purely Jewish event. In this sense, the ADL willingly diluted much of the moral or political capital it would have gained by pushing the Holocaust into the American consciousness. In addition, many of the first teachers of the Holocaust were not Jewish, and they taught in areas with marginal Jewish populations. These conclusions complicate any claims that the Holocaust was pushed into the American consciousness by influential Jewish leaders, or, at the very least, they recognize that the positive reception of the Holocaust by non-Jewish Americans was a crucial component of its popularity.

The success of the Holocaust educational movement also complicates the assertion that the reforms of the new social studies and "affective revolution" were misguided failures. Indeed many of the curricular materials and ideas introduced and imposed in the 1960s and 1970s were not wholeheartedly adopted, and these hastily implemented reforms certainly contributed to the state of confusion that characterized the social studies curriculum in the late 1970s. But these very same reforms created a looser curricular environment at the secondary level in which teachers could experiment with new ideas and materials. Holocaust education was the result of this curricular freedom. The popularity of these specific curricula enabled Holocaust education to survive into the 1980s and 1990s, even when the general public had turned against the neo-progressive curricular reforms of 1960s. The continued ascendancy of Holocaust education in the age of back-to-basics suggests that the new social studies and affective revolution were not complete failures.

Chapter 6

Critiquing Holocaust Education

If the 1978 NBC *Holocaust* miniseries helped spark mainstream interest in the event, then 1993—the year in which Stephen Spielberg's *Schindler's List* and the United States Holocaust Memorial Museum (USHMM) opened to critical acclaim—represented the culmination of this interest. During the years leading up to 1993 there was a steady rise in educational interest in the Holocaust across the country. This was accompanied by a proliferation of Holocaust educational material. By 1993, Holocaust education was so established that the national Museum did not feel the need to create an "official" Holocaust curriculum. Instead, under the leadership of William Parsons and Samuel Totten, the USHMM issued a set of teaching guidelines meant to direct teachers in the selection and refinement of existing curricula. Many of these guidelines were critical in nature, indirectly attacking the quality and effectiveness of many of the units covered in this study. The guidelines suggested that Holocaust education, as a movement, had moved beyond justification and implementation to all-out critique. The continued rise in interest about teaching the event was underscored by the educational campaign launched in coordination with the screening of *Schindler's List*.

Schindler's List

Over 120 million Americans viewed Spielberg's academy-award-winning film, which related the story of profiteer-turned-rescuer Oskar Schindler, played by Liam Neeson. Based on a true story, the film depicted how Schindler exploited the Nazi's rise to power by using flattery and bribes to win military contracts. Over the course of the film he befriends a Jewish accountant and financier Itzhak Stern, played by Ben Kingsley, to help run the factory, which Schindler staffs with unpaid Jews from the Krakow ghetto. After all of Krakow's Jews are assigned to the Plaszow Forced Labor Camp, overseen by the ruthless Commandant Amon Goeth, played by the Ralph Fiennes, Schindler arranges to continue using Polish Jews in his plant. When he sees what happens to many of his employees in the camp, he develops a conscience and realizes that his factory is the only thing preventing his staff from being shipped to the death camps. Schindler then demands more workers and using his entire personal fortune bribes Nazi leaders to keep Jews on his employee lists and out of the camps. In this manner, Schindler ultimately saves 1,100 Jews.

While not as historically comprehensive as the NBC miniseries, Spielberg's film included graphic images of concentration camps, gas chambers, Jewish transports, ghetto liquidation, and numerous examples of Nazi cruelty and indifference. Most film critics offered high praise for the artistry of the directing and sensitivity shown toward the subject matter. Some critics were concerned that the director of *E.T.* and *Jurassic Park* was going to present the Holocaust as a suspense-filled, special-effects extravaganza. They were pleasantly surprised to find the film to be an immensely moving masterpiece. In fact, they specifically praised Spielberg's ability to reign in his dramatic tendencies, and lauded his range of impressive new techniques. Predictably, however, academics expressed a host of problems with the film.

For certain scholars, the emphasis on the heroism and righteousness of Schindler obscured the fact that the majority of the Germans were either perpetrators or bystanders; Schindler's story was the exception, not the rule. Others objected to the fact that the film had few major Jewish characters. The Jewish victims were depicted namelessly from the German point of view. A more significant critique was that the film relied on Hollywood conventions such as melodrama, emotionally manipulative music, and, worse of all, a happy ending. Critics called the act of placing a redemptive ending on the event—whether it be the founding of Israel or the moral conversion of Oskar Schindler—the "Americanization" of the Holocaust. Such an accusation would also be hurled inappropriately at certain Holocaust curricula that focused too heavily on the universal aspects of the event or devoted too much time to rescuers and heroic acts. Finally, particularist critics objected to the entire idea of using a film to represent the Holocaust in any accurate sense. Such a task was not only impossible, but also morally wrong. The thought of Spielberg and his staff pouring over the choice of the correct wardrobe and makeup to "capture" the reality of death was offensive. "What on earth did they think they were doing?" Leon Wieselter wondered, "Do they really think they got it right?"[1]

Spielberg's campaign to use *Schindler's List* to spread awareness about the Holocaust involved a coordinated educational effort far more involved than the one engineered to support NBC's 1978 *Holocaust* miniseries. He screened the film for free for almost two million high-school students in the spring of 1994 in more than forty states. He requested Facing History Ourselves Foundation to prepare a study guide to accompany the film, which—after it was no longer available in theatres—he made available as a videocassette to every high school in the country. The 654-page study guide was divided into three parts. The first centered on the historical context surrounding Oskar Schindler. The second included discussion questions about the film, including comments by *Schindler's List* author Thomas Keneally and Spielberg. The third section, true to the tradition of *Facing History*, investigated the universal themes and moral implications of the film.[2]

After he was done promoting the film, Spielberg also created the Survivors of the Shoah Visual History Foundation, which documented the testimonies of thousands of Holocaust survivors. Spielberg selected Micheal Berenbaum, a young Jewish scholar who served as deputy director for the president's Holocaust Commission, to head the foundation. Berenbaum hoped to reach out to all existing Holocaust educational organizations to make the testimonies available. "We want a curriculum," Berenbaum explained, "for Roman Catholic education, for Orthodox Jewish education, and for secular education. Each of these types of schools wrestles with different issues and different values that concern them."[3] Through this organization Spielberg hoped to encourage Holocaust education long after *Schindler's List* had faded from memory. He agreed with Berenbaum that the way to keep the event in the American consciousness was to make it relevant to as many groups as possible.

Schindler's List was also popular abroad. In the United Kingdom, the Holocaust Educational Trust sent an edited copy of the film to every secondary school in the country free of charge. Just as in America with the concurrent opening of the United States Holocaust Memorial Museum, in Britain the film also corresponded with an event that helped to spread Holocaust education throughout its schools. In 1988, the British government passed the Education Reform Act that set out to establish strict guidelines for a national curriculum. Originally, the National Curriculum History Working Group left out the rise and fall of Nazi Germany, explaining that it did not mean "to downplay the importance of these or other events, but for every suggested addition, something has to make way."[4] After substantial objection and protest by teachers and academics, the

History Working Group included the Holocaust explicitly in the revised curriculum published in 1990.

Unlike the localized American educational system, the centralized nature of the British system allowed them to expand the teaching of the Holocaust almost instantaneously, despite the fact that many British teachers were not knowledgeable or qualified to teach about the event. Just as in the United States, British history textbooks only offered superficial coverage of the event. To remedy the situation, Carrie Supple published a supplementary text titled *From Prejudice to Genocide*. It was in this context that *Schindler's List* was first distributed to schools, where it would become the most widely used Holocaust video in the United Kingdom.[5]

On the heels of Spielberg's film, in 1995 the National Council for the Social Studies (NCSS) devoted another issue to Holocaust education. The publication included several informative essays on Holocaust denial, non-Jewish victims, German women, and anti-Semitism, as well as teaching ideas on how to incorporate primary sources, poetry, and literature. The guest editors, Samuel Totten and Stephen Feinberg, interviewed Spielberg, who declared that his "primary purpose in making *Schindler's List* was education." Spielberg shared a story of a screening at the Apollo Theatre in New York. A student there (presumably black) expressed that the Holocaust wasn't his story, to which another assertively replied, "pain is pain." Teachers, Spielberg explained, should "make the study of the Holocaust and issues of hatred and intolerance as relevant as possible so that it can have a real meaning and impact for every individual in their own lives." He hoped that the film would serve as a "door toward a wider teaching of tolerance, covering slavery, the Native-American History, the immigration story, and wide base of ethnic, religious, and gender issues." Spielberg was a strong and very public supporter of universalizing the Holocaust. His film addressed racial hatred, he explained, "You can either say it is just a Jewish issue and walk away from it, or you say it is about racial intolerance everywhere in the world. When you can look at it broad-mindedly, then it isn't just about the Jews."[6] As we shall see later, it was this kind of "broad-minded" thinking that the USHMM hoped to reign in with its exhibit and educational program. For critics, Berenbaum and Spielberg become the twin pillars of the Americanization of the Holocaust in the 1990s.

The United States Holocaust Memorial Council

While Holocaust education was expanding during the 1980s, the United States Holocaust Memorial Council was working on the design of the new

U.S. Holocaust Memorial Museum (USHMM). Like the curriculum designers in this study, the Holocaust Council also had to navigate between the Scylla of making the Holocaust relevant to all Americans and Charybdis of respecting the particularities of the Jewish experience. But, the design of the museum differed from the writing of Holocaust curricula in two fundamental ways. First, the USHMM represented the endorsement and funding of the federal government. The Council's position essentially presented the "official" U.S. interpretation of the Holocaust to the nation and to the world. Therefore, the Council could not shrug off the objections of political constituencies, ethnic interest groups, and Holocaust scholars as easily as the teachers in the previous chapters had done. Second, the museum's exhibit represented a permanent, fixed interpretation of the event. Each visitor, regardless of his/her individual background, would experience the same narrative of the event. While different people would bring their own perspectives to the exhibit, the arrangement and explanation of the artifacts themselves were permanent; it was not flexible like a Holocaust curriculum, in which teachers could make subtle or major changes to fit the needs of their individual classes. Therefore, the museum designers felt compelled to plan the exhibit carefully. Like a textbook, the museum would need to be as culturally neutral as possible, in a way that would offend the least amount of people. Although pedagogy would play a part in the overall design, the Council's primary concerns were political and cultural in nature. They were indeed presenting an official, "Americanized" version of the Holocaust.

In the summer of 1978, Elie Wiesel received a phone call from Stu Eizenstat, President Carter's advisor of internal affairs, to inquire about why Wiesel had not been returning his phone calls. Wiesel, thinking that the messages he had received to call the president of the United States had been a joke, simply ignored his previous attempts to contact him. When Eizenstat finally reached Wiesel, he told him that the president had selected the author to chair the newly formed President's Commission on the Holocaust. Initially, Wiesel refused. First of all, Wiesel insisted that as a Jew he was opposed to monuments and memorials. Second, he did not believe that the Holocaust could be represented adequately and respectfully. "How are we going to 'show,'" he wondered, "when it is almost impossible to speak of it?" But most of all, he did not want to become entangled in the cultural politics that would inevitable ensue. Nonetheless, after a meeting with the President Carter himself, he was persuaded to accept the position. With Wiesel at the helm, the Commission consisted of thirty-four members including senators, congressmen, survivors, religious leaders, and scholars.[7] They issued their *Report to President* in September 1979.

"From the beginning," Wiesel recalled, the meetings were "dominated by the question of the specificity and universality of the Holocaust."

Throughout the process Wiesel consistently guarded his views of the Holocaust as a "Jewish tragedy with universal implications."[8] As representatives from the other persecuted minorities questioned Wiesel's interpretation, he stood firm in his insistence of a Jewish core to the definition of the event. He cast himself as the guardian of the Jewish Holocaust victims, whose memory was under assault by those who wanted to stretch the boundaries of the event to meet the needs of the nation. Initially, the issue was to what degree the definitional uniqueness of the Holocaust would be respected? During his speech for the Holocaust Commemoration ceremony in April 1979, President Carter shocked Wiesel by asserting that the Holocaust referred to "11 million innocent victims exterminated—6 million of them Jews," thereby, denying the definitional uniqueness of the Jewish experience.[9] After the ceremony Wiesel confronted the president and politely expressed his objection to this number. As he later explained: "It's true; there were others as well. So they said 11 million, 6 million of whom are Jews. If this goes on, the next step will be 11, including 6, and in a couple of years, they won't even speak of the 6. They will speak only of the 11 million. See the progression? 6 million plus 5, then 11 including 6, then only 11."[10]

Some criticized Wiesel's view arguing that "all Slavs of Eastern Europe and Russia were slated for decimation, degradation and eventual liquidation," and that it was "morally repugnant to create a category of second-class victims."[11] Others insisted that the new museum established the Holocaust as an American experience, not just a Jewish one. Therefore, the Jewish community would have to make accommodations to "Americanize" the exhibit. In the words of Michael Berenbaum, the museum would need to "resonate not only with the survivor in New York and his children, but with a black leader from Atlanta, a Midwestern farmer, or Northeastern industrialist."[12] Wiesel insisted that Americans could take these universal aspects from an exhibit that exclusively covered the Jewish experience.

In 1979, the president transformed the Holocaust Committee into the U.S. Holocaust Memorial Council with Wiesel as its chairman. As the planning progressed, the cultural politics, backroom deals, and constant bickering were too much for Wiesel to endure. He was consumed for hours a day on the phone with the various Council members trying to assuage the unraveling problems while stubbornly defending his own views. In December 1986, despite the pleas of the other Council members, an exhausted and frustrated Wiesel resigned from the chairmanship. At the time he explained to the Council that as the project moved to the next phase, he was stepping down because he knew very little about finances and management. He later revealed that he was upset with the remarks of President Reagan, who at a military ceremony in Bitburg, Germany, in

1985 asserted that the SS troops killed during World War II "were victims just as surely as the victims in the concentration camps." The president's comments incited a controversy that had, once again, thrust the Holocaust onto the front pages of newspapers. Wiesel later reflected, "How could I 'serve' under a president who 'objectively' has whitewashed the SS by comparing them to their victims?"[13] Ultimately, with Wiesel's resignation, the Council entered its more productive stage, and completed the design of the exhibit for its opening in 1993.[14] But Wiesel's early influence as Council chair ensured that the Museum would retain the Jewish uniqueness of the event.

The 1979 *Report to President* also suggested that the museum should be accompanied by an educational foundation to "assist with the development of appropriate curricula and resource material while working cooperatively with those school systems which wish to implement the study of the Holocaust."[15] The Commission hoped that the study of the Holocaust would become "part of the curriculum in every school." Initially the educational foundation was placed on the backburner, but as the nature of the exhibit became more concrete, the Museum planners began to concentrate on its educational components. William Parsons joined the "bigger classroom" of the Museum in 1991 as director of education. Parsons had parted ways with Facing History in 1987 because he felt that the foundation had been drifting from his original intentions. He had hoped that Facing History would serve as a national center for genocide (past and present) study and education, for which he felt there was a pressing need.[16] The Holocaust Council, on the other hand, had a far more specific objective regarding the Holocaust. When Parsons joined, he was given an official "Mission Statement" on which to base his curricular materials. The definition of the Holocaust, which reflected Wiesel's enduring influence, read as follows:

> The Holocaust was the state-sponsored, systematic persecution and annihilation of European Jewry by Nazi Germany and its collaborators between 1933 and 1945. Jews were the primary victims-six million were murdered; Gypsies, the handicapped, and Poles were also targeted for destruction or decimation for racial, ethnic, or national reasons. Millions more, including homosexuals, Jehovah's Witnesses, Soviet prisoners of war, and political dissidents, also suffered grievous oppression and death under Nazi tyranny.[17]

Unlike Facing History, the museum was to concentrate on the "unprecedented tragedy" of the Holocaust. It was not to investigate other genocides or compare the Holocaust to recent or contemporary events. Nonetheless, Parsons enthusiastically accepted the position.

USHMM's Guidelines for Teaching about the Holocaust

Parsons insisted that the Museum should not issue an official curriculum, but instead it should develop a set of teaching guidelines. Since he obviously identified with the many innovative teachers around the country who had designed and implemented their own Holocaust curricula, he did not want "to step on any toes." Instead he wanted the USHMM to serve as a clearinghouse for the networking of Holocaust educators and dissemination of Holocaust materials.[18] Parsons organized a team of educational consultants to begin work on several Holocaust-education-related projects, including a poster series based on exhibit artifacts, and a series of national teacher workshops. The guidelines were published in 1993 to correspond with the opening of the Museum. They were coauthored by Parsons and Samuel Totten, who had worked as an unpaid volunteer under Berenbaum in the summer of 1979.[19]

Totten had enthusiastically joined the planning for the Museum as an educational consultant in late 1980s. His background was notably different from the other members, who had either been pioneers in Holocaust education or Holocaust history experts. Totten had grown up in Los Angeles in a home with a racist, abusive father. "A day did not go by," he recalled from his childhood, "that either my mother, brother, or I was not screamed at, slapped, punched, pounded, bitten, or hit with whatever was in reach." Understandably, his brutal childhood left a lasting impression, instilling him with a "marrow-deep disdain for those who brutalize others" and a lifelong interest in the protection of human rights. During the 1970s, Totten taught English and worked with Amnesty International in Australia, Nepal, and Israel. While teaching in Jerusalem, he visited Yad Vashem, the Holocaust Martyr's and Heroes Remembrance Authority, and he was introduced to the literature of Elie Wiesel, where he gained a deeper understanding of the horrific details of the Holocaust. When he returned to the States in 1978, he tenaciously wrote several members of the Holocaust Museum Education Committee expressing his hopes of becoming involved with their work. At that stage, they did not need his assistance full time, so Totten accepted a teaching job in California, where he added Holocaust literature to his course reading list. From 1980 to 1985, he attended Teachers College (Columbia University) in New York where he received a doctorate in Curriculum and Instruction. He had hoped to do his dissertation on the focus, methods, and efficacy of Parson and Stern Strom's *Facing History* curriculum, but instead he was persuaded to research educational issues surrounding the nuclear arms race. After he graduated, he accepted a position at the University of Arkansas-Fayetteville. In 1989, he was asked to join the team of educational consultants for the Museum.[20]

Totten had first met William Parsons at a NCSS conference in Dallas in 1987, where they both participated in a seminar on human rights. In the late 1980s, they coedited a special issue of *Social Education* on "Teaching about Genocide." Like Parsons, Totten was interested in the broader issue of genocide, but he also had an informed respect for the uniqueness of the Jewish experience during the Holocaust. They both understood the complexity of history and were dedicated to conveying the facts to students in a historically accurate manner. They also believed in the progressive educational ideas of the social studies—particularly regarding the ineffectiveness of the traditional approach to teaching history. Totten recalled their similar pedagogical outlook:

> Though we did not use such terminology at the time, we both practiced a constructivist type of pedagogy in which we developed strategies and activities whereby students would wrestle with key ideas, concepts, events and come to a deep understanding of them, through deep thought, written reflection, discussion and debate. At one and the same time, we both looked askance at "direct instruction" whereby a teacher pours huge amounts of facts into students' heads and then expects the students to spit it back out on tests. Equally significant, we are both firm believers in depth over coverage, and that too drew us together.[21]

Despite their similar outlooks, the two authors did diverge on a few issues. Totten wanted the guidelines to target directly what he considered the poor quality of many of the Holocaust curricula in use. Over the years he had closely examined many of the more acclaimed and popular Holocaust curricula and concluded that "many were rife with historical inaccuracies... were often pedagogically unsound... [and] many of the learning activities in the lessons were set at the lowest levels of the cognitive domain."[22] Parsons, on the other hand, wanted to be more politically correct in his assessments by using "softer language" to suggest these shortcomings in a more subtle manner. Totten also wanted to bring more attention to issues of contemporary human rights violations and other twentieth-century genocides but, as he explained, such ambitions "got scrapped" early in the process.[23]

As educational chair, Parsons was in more of a political position and felt that he had to adhere to the "Mission Statement" that had been so painstakingly negotiated by the larger Council. Therefore, he thought the guidelines should only address the Holocaust. Parsons also seemed to be moving in more of a traditional direction, regarding his pedagogical orientation to teaching history. While he had codesigned the *Facing History* curriculum in the late 1970s from a progressive perspective, he was becoming more interested in the historical particularities of each genocidal event. He left the Foundation in 1987, when he thought that it was adopting too extreme of a universalist approach.

Parsons worried that the historical particularities of the individual events were being compromised in an attempt to make them applicable to contemporary issues. His experience with the Museum confirmed his ideas. Through his interaction with the many Museum staff historians and the historical artifacts that made up the exhibit, Parsons gained an even greater respect for the historical particularities of the Jewish experience. He wanted to convey the uniqueness of the Holocaust in the guidelines in way that would not diminish the suffering of other historical victims. Therefore, as Totten and Parsons approached the guidelines, they both had different agendas. Totten wanted to critique, as bluntly as he could, the "weak and sloppy curricula and resources that were, at the time, available for teachers." Parsons wanted to emphasize the historical particularities of the Jewish experience in the Holocaust as a way to complicate simple conclusions.[24] Both their concerns were represented.

The guidelines were originally issued in 1993 as a pamphlet. In 1995, they were combined with other USHMM teaching resources into a book called *Teaching about the Holocaust: A Resource Book for Educators.* The first guideline insisted that teachers "Define the Term Holocaust" and provided, word for word, the official definition provided in the "Mission Statement." The next guideline attempted to correct some of the potential shortcoming of the progressive approach to teaching the Holocaust by suggesting that teachers avoid comparisons of pain. "One cannot assume," the authors wrote, "that the horror of an individual, family, or community destroyed by the Nazis was any greater than that experienced by victims of other genocides." They stressed that each genocidal situation was unique and that teachers should not enter into any exercises that involved creating a hierarchy of suffering. This also implied that the uniqueness of the Holocaust presided in the unprecedented nature of the Nazis institutional–political coordination, not in the individual experiences of those who lived through the persecution. Or, as the authors explained, one should not assume that "the victims of the Holocaust suffered the most cruelty ever faced by a people in the history of humanity."[25]

The next three guidelines mostly reflected the views of Parsons, because they emphasized that teachers should pay close attention to the historical particularities of the Holocaust. The suggestion to "Avoid simple answers to complex history" was aimed at those curricula, such as the Roselle Chartock's *Society on Trial* and Richard Flaim and Edwin Reynolds' New Jersey unit, which employed a progressive, affective orientation that often subsumed all prejudice under the heading of racism. "Racism has a cohesiveness to it," one teacher proclaimed in a 1988 *New York Times* article on Holocaust education, "The same kind of mentality keeps it going, whether you're prejudiced against Vietnamese, Puerto Ricans, blacks or Jews."[26] This was the type of generalization to which the authors objected. "The Holocaust was not simply the logical and inevitable consequence of unbridled racism," they explained in the guidelines, "Rather, racism combined with centuries-old bigotry and anti-Semitism; renewed by a nationalistic fervor that emerged in Europe in the latter half of the nineteenth century." The authors also suggested that teachers should avoid the impression that the Holocaust was inevitable, and insisted that they "strive for precision of language" when referring to such issues as concentration camps and resistance.

Overall, the guidelines emphasized more of a disciplinary approach to teaching history than any of the previous curricula had done. A disciplinary approach taught students how to "be historians" by engaging directly with primary sources and trying to reconcile conflicting accounts. To avoid indoctrination, teachers would not "transmit" any historical facts, instead allowing the students to construct their own knowledge through interaction with the sources. While Parsons and Totten did not go this far, they did suggest a lot of disciplinary-like methods. One guideline suggested that students should make "careful distinctions about sources of information... students should be encouraged to consider why a particular text was written, who wrote it, who the intended audience was, whether there were any biases inherent in the information, [or] whether any gaps occurred in the discussion." The authors encouraged the liberal use of first-person accounts and primary sources. In addition, they emphasized the importance of historical context in explaining the motivation of historical actors. This meant de-emphasizing the elements of the Holocaust that are comparable to other historical events and contemporary issues. Accordingly, the guideline's overall outline for teaching the Holocaust did not begin with a discussion of historical precedents to the Holocaust, nor did it suggest a behavioral investigation into the nature of man, as most of the previous curricula had done. Instead it suggested a chronological progression through the various events. They wanted students to think critically about the complexity of "real" history. Parsons, in particular, had always centered his instruction on "complicating" students thinking.[27]

Part of the reason for this renewed interest in the disciplinary approach derived from an implicit critique of previous curricula, as well as the USHMM's mandate to respect the uniqueness of the event. But there was also an ongoing revival in the disciplinary approach in field itself. In the late 1980s and early 1990s, academic historians like Gary Nash, Diane

Ravitch, and Paul Gagnon had been mobilizing support for reasserting the centrality of history in the school curriculum. Riding on the wave of "back-to-basics," these scholars championed the importance of real history (as opposed to social studies) and argued for a revival in the traditional orientation to teaching the past. As part of this movement—and to a large degree, in response to it—cognitive psychologist Sam Wineburg published a series of influential articles on the cognitive aspects of "historical thinking," which further highlighted the complexity of the historian's work and his/her interaction with primary sources. This renewed interest in "thinking historically" resonated with educators who were interested in using history to develop critical thinking skills in students. Thus, the suggestions of the guidelines reflected the overall curricular zeitgeist, which was moving away from the affective approach to teaching history toward a more disciplinary one. Of course, such generalizations cannot be taken too far, since, as we have seen in the last chapter, progressive curricula like *Facing History* continued to be popular.[28] The point is that the emerging research on history education supported many of the ideas put forth by the guidelines.

The Museum guidelines also included some of Totten's pointed criticisms dressed in the guise of suggestions. The first of these was an attack on those Holocaust accounts, such as Bea Stadler's *The Holocaust: A History of Courage and Resistance* textbook, which stressed the positive elements of the tragedy. "[A]n over emphasis on heroic tales in a unit," the authors admonished, "can result in an inaccurate and unbalanced account of the history." Next Totten targeted those curricula that used affective shock treatments to inspire students' interest, such as Chartock's *Society on Trial.* "Graphic material should be used judiciously," they explained, "try to select images and texts that do not exploit the students' emotional vulnerability or that might be construed as disrespectful of the victims themselves." Finally Totten attacked those teaching methods that he considered inappropriate and pedagogically unsound, such as "word scrambles, crossword puzzles, and other gimmicky exercises" that tend to "trivialize the history." Of course, no other technique allegedly trivialized the Holocaust more than Rabbi Zwerin's *Gestapo* game. In this case, the objection was clearly articulated:

> In studying complex human behavior, many teachers rely upon simulation exercises to help students "experience" unfamiliar situations. Even when great care is taken to prepare a class for such an activity, simulating experiences from the Holocaust remains pedagogically unsound. The activity may engage students, but they often forget the purpose of the lesson and, even worse, they are left with the impression at the conclusion of the activity

> that they know what it was like during the Holocaust... It is virtually impossible to simulate accurately what it was like on a daily basis with fear, hunger, disease, unfathomable loss, and the unrelenting threat of abject brutality and death.[29]

The rest of the resource book supported the guidelines by providing an abridged but thorough narrative of the Holocaust with a detailed chronology. It ended with an extensive annotated bibliography and videography.

Like the previous curricula designers had done, Parsons and Totten had designed the resource book to be flexible enough to allow teachers to approach the event in a manner they deemed fit, but useful enough to actually assist them. Rather than listing excerpts from hundred of historical and social scientific sources, the USHMM resource book's annotated list described the contents of scholarly Holocaust monographs and celebrated Holocaust memoirs. In this way, teachers would be encouraged to go directly to the sources themselves, which would naturally lead them to appreciate further the historical particularities of the event. This was another subtle way to endorse a disciplinary approach.

So what caused Parsons and Totten to become such ardent protectors of the historical, definitional, and, to a certain extent, metaphysical uniqueness of the Holocaust? Neither was Jewish, and both had a strong allegiance to universal concerns such as contemporary human rights violations and genocides. Intuitively, given their backgrounds, one would think that they would want to Americanize the event. But this wasn't the case. Parsons and Totten both felt a responsibility to respect the exceptionalism of the Holocaust, not for cultural reasons, but for pedagogical ones. Not because they felt pressured by Jewish influences (although this was an influence on Parsons), but because they believed that emphasizing the uniqueness of the Holocaust was the most effective way to achieve their pedagogical objectives; the best way to inspire critical thinking and to complicate students' assumptions. A universalized Holocaust that subsumed all genocidal acts under the heading of racism or viewed the Holocaust through an investigation into human nature oversimplified the complexity of the event. The authors designed the guidelines to make this very point.

The guidelines clearly displayed elements of the traditional, progressive, affective, and even (in the assertion that a simulation inherently trivialized the event) particularist approaches to history. But overall it emphasized the disciplinary approach. While cultural influences cannot be completely discounted, I have suggested that this subtle turn away from affective, progressive, and behavioral approaches toward a more disciplinary orientation reflected greater changes in the field of history education.

This shift continued to affect Holocaust education in the years that followed. In 2002, Samuel Totten and social studies educator Karen Riley collaborated on an article that, using the Museum guidelines and the "historical emphathy" instructional model, critically assessed Holocaust curricula for Connecticut, Virginia, Florida, and California. Historical empathy was a new term employed by researchers to describe the goal of imparting the complexity of history in the minds of students.[30] In their research Totten and Riley discovered that these curricula failed to address historical empathy adequately and violated many of Totten's guidelines by displaying a "simplistic portrayal of complex history, a lack of adequate information and/or omissions of key issues and events, and watered-down concepts."[31]

Totten and Riley's article demonstrated how, in the new century, the goals of Jewish elites and certain social studies educators have, to a large degree, become aligned. Both groups endorse a respect for the exceptionalism of the Holocaust. Both groups want students to appreciate the particularities of the Jewish experience. But they wanted these things for different reasons. Certain Jewish scholars want the Holocaust to be considered unique out of respect for the victims and for its religious significance. Certain educators want the Holocaust to be considered unique because it is a complex phenomenon that can create cognitive dissonance in the minds of students. Such critical thinking skills are required, educators argue, to function as an active citizen in a democracy. Admittedly, the distinction between these two rationales is often hard to make. As we shall see later, they often overlap. Nevertheless, scholars should not interpret the recent inclination toward the particularist stance on the Holocaust as a victory by Jewish elites, but rather as a temporary confluence of orientations.

Holocaust Educators on the Offensive

After the USHMM teaching guidelines were published and distributed to thousands of schools across the country, Parsons continued to work for the Museum and Totten returned to the University of Arkansas. While Totten's critical eye was contained by the political nature of the Museum, as a professor, he could write as feely as he wished. For the next decade, Totten published numerous articles critiquing existing Holocaust curricula and attacking "unsound" teaching approaches such a Holocaust simulations and teaching the event to young children. He even criticized a set of guidelines for evaluating Holocaust curricula that had been drafted by the

Association of Holocaust Organizations. Through this work, Totten has become the most prolific (and controversial) writer on Holocaust education in the country.[32]

Totten began his pedagogical interest in the Holocaust from the perspective of literature. At first he felt compelled to teach the event solely through the classic books, but since then his ideas have "changed radically." By the 1990s, he had developed more of a disciplinary, historical stance. He recommended using literature "to help illuminate certain issues that history raises," but considered himself "more in line with how a historians would go about teaching about the Holocaust." His approach to the studying of the event and to writing about it were influenced by his graduate work at Teachers College, Columbia University, where he studied under educational historian Lawrence Cremin and educational philosopher Maxine Green. These scholars inspired Totten "to appreciate the need to look below the surface in order to scratch away the given, official story." Green in particular pushed Totten "to constantly posit penetrating questions, to get at the 'why' behind the 'what' and 'how' of education and life, and to realize that we all have blinders... that we have to push and prod our way through."[33]

As mentioned earlier, Totten felt restrained by the USHMM. He had hoped to use the guidelines to take strong, clear stands against certain activities, especially Holocaust simulations. In 2000, he published an article in *Social Education* arguing that the use of simulations constituted "poor pedagogy as a result of drastic over-simplification of Holocaust history." Oddly, his argument drew little upon pedagogical theory, and instead presented an argument that veered close to Elie Wiesel's particularist perspective. "To take something so profoundly disturbing, and so overwhelming to those who lived through it," he explained, "and to turn it into something that becomes for many... 'fun and games' is to make a mockery of what the victims lived through." He found it "galling" to think that teachers could actually provide their students with a real sense of what it was like or even approximate the experience.

To support his argument, Totten quoted an excerpt from Elie Wiesel asserting that the Holocaust is "not a subject like all the others. It imposes certain limits... [and] demands a special sensibility, a different approach, a rigor, strengthened by respect and reverence and above all, faithfulness to memory." In the article Totten presented several examples of eyewitness accounts of horrific atrocities performed during the Holocaust as justification for its uniqueness. Even though the USHMM guidelines insisted that teachers should avoid creating a hierarchy of suffering, Totten seemed to be doing just that. He selected experts from testimonies depicting the very events teachers sought to replicate in their classrooms. His objective seemed to be to convince teachers that these events were too horrible to be replicated. In the place of simulations, Totten suggested that students engage in the reading of primary source accounts. In this way, his advice was consistent with his turn away from the literary toward the historical approach, away from the affective and artistic aspects of fictional accounts toward the factual, evidence-based testimonies of the victims.[34]

Totten also took a firm position against teaching the Holocaust to elementary school students below the fourth grade. He directly targeted an article by Harriet Sepinwall that appeared in the NCSS journal *Social Studies and the Young Learner,* suggesting strategies for teaching the Holocaust to primary age children. Totten found such an idea "problematic." He argued that what Sepinwall described and advocated was "not so much Holocaust education as prejudice reduction, bias reduction, or conflict resolution," which were all admirable goals, but do not address "much of anything about the history of the Holocaust." Totten suggested that teachers stop using the term Holocaust education, and instead use "Pre-Holocaust education" or "Preparatory Holocaust Education."[35] Like his critique of simulations, Totten did not draw upon any cognitive or developmental research to support his position, but instead drew upon specific facts and testimonies of the Holocaust itself. Again, he seemed to be setting limits around how the Holocaust can and should be approached—a process that seemed to reify the act of teaching the event, while at the same time mystifying it as history. To Totten and others, the Holocaust needed a unique set of pedagogical rules.

After publishing this critical work, Totten reported, he received "all sorts of reactions from Holocaust scholars and educators, ranging from kudos (someone had to eventually say it, good job) to comments that I was mean-spirited, overly critical and harsh."[36] He was not alone in his critique of Holocaust curricula. In general, by the 1990s, a new consensus emerged among social studies educators that those curricula, like *Facing History,* that employed the affective, behavioral approach were inadequate and ill-conceived. As Totten explained, "I believe that an in-depth, chronological examination of the history... whereas *Facing History* takes a different approach."[37] He agreed with most of criticisms voiced by Deborah Lipstadt in her 1995 review of the popular curriculum. Likewise, in a 2001 critique Karen Riley, a professor of education at Auburn University, suggested that the purpose of the *Facing History* curriculum was to "use the Holocaust as a platform for teaching moral behavior and shaping attitudes, rather than to help students acquire an understanding of the Holocaust as an historical event."[38] *Facing History* continued to be the most popular curriculum in the county and, therefore, continued to draw the most attention.

In addition to *Facing History*, Riley also voiced criticisms of Leatrice Rabinsky and Carol Danks' *The Holocaust: Prejudice Unleashed* and the Michigan curriculum *Life Unworthy of Life: A Holocaust Curriculum*. The former, she argued, presented simplistic historical depictions. The curriculum instructed students that the Nazis had carefully developed the final solution ahead of time, and that the abandonment of the Jews could be attributed solely to apathy instead of German nationalism and racism. The latter curriculum, she suggested, relied upon a simplistic "inoculation rationale" that "humankind is doomed to repeat itself." As a result students are "lulled into a certain comfort that the right questions have been asked and answered and that all that is required is to execute a few activities so that the correct interpretation is enforced." Overall, Riley argued, both these curricula focused the study of the Holocaust on the transmission of certain emotions, facts, and lessons. Thus, the Holocaust was reduced to a series of correct answers for students to learn, instead of approaching the topic through open-ended inquiry.[39] Both Totten and Riley agreed that inquiries into the Holocaust must draw upon the discipline of history. Like historians, students should pose difficult questions, consult multiple sources, and construct evidence-based answers. Such activities would develop students' historical empathy.

A 1995 oft-cited article by Karen Shawn, director of Holocaust Studies for the Moriah School of Englewood, New Jersey, presented an even more pessimistic view of the state of Holocaust education in American schools. She worried that the rapid, broad-based popularization of the event, exemplified by the mandating of the genocide education in certain states, was diluting and diminishing its impact. She argued that legislators might have put the cart before the horse by mandating the teaching of the event, before providing training for teachers in how to do so. The vast majority of teachers, she declared, "through no fault of their own, currently lack the basic skills necessary to implement state mandates with professional integrity." She suggested that state governing bodies establish minimum teaching standards for Holocaust education requiring them to pass certain courses of study. Teachers who complete the required course should be certified to teach the event, and those who do not, she argued, "should not be teaching it."

The "ill-considered rush" to educate all students about the Holocaust, Shawn complained, was leading to "a recent alarming proliferation of poorly conceived and executed textbooks, teaching aids, and lessons plans flooding out schools." Among these "recent" curricula she listed Rabbi Zwerin's *Gestapo* simulation board game, which she found "profoundly disturbing." The loss of historical integrity inherent in *Gestapo*, she mockingly remarked, "is small price to pay to keep the peace in a classroom filled with disaffected teens." Like Totten, she was appalled by teacher's use of multiple-choice tests, crossword puzzles, and word searches, which allegedly trivialized the event.[40]

Similarly, Stephen Wylen, a New Jersey rabbi, complained in a local newspaper editorial that the Holocaust curriculum in his state was "so watered down that the point is entirely lost." He particularly objected to the use of the *Diary of Anne Frank* as a Holocaust text. He suggested that teachers use Wiesel's *Night* in its place. A real Holocaust curriculum, he explained, must confront the history of anti-Semitism, the reality of mass murder, and the dark and evil side of the human psyche.[41] Anne Frank's testimony does not do this.

Over the course of 1990s and 2000s, several Holocaust educators offered advice on how to teach the event in secondary schools, much of it drawing upon their own experience with college-age students. Samuel Totten published a book of Holocaust education resources and issues in 2003. He offered specific advice on how to construct a rationale for teaching event, how to assess and work with student's prior knowledge, how to get students to think critically about the event, and how to close the unit effectively.

Totten also included a chapter on common misconceptions about the Holocaust, which included that the event was inevitable, it could be boiled down to only one or two major causes, the victims somehow brought the persecution upon themselves, Jews went like sheep to slaughter, Jewish resistance was only physical in nature, death camps and concentrations camps were the same, all camps were located in Germany, only Jews were targeted by the Germans, there is an "Aryan Race," and Nazis who refused to kill Jews were themselves murdered. Totten devoted an entire chapter to the misconception that Jews were a race, a concept he insisted "Educators Must Get Right."[42] The book was filled with lengthy quotations from numerous historical works on the Holocaust and it concluded with an extensive annotated bibliography. As an astute reviewer wrote about the book, "To expect a secondary teacher, who frequently must cover the entire span of history from the 'big bang' to the day before yesterday on one or two semesters, to digest this awesome amount of material seems naïve."[43]

By now the reader is probably experiencing Holocaust education criticism fatigue. My narrative has traced nearly a half-century of scholars complaining about the poor quality of Holocaust education in American public schools. Many professors, who may or may not have ever stepped foot in a public school classroom, have written numerous essays demanding that Holocaust education be improved. Likely many more will do so in the future. Although their scholarly criticisms are mostly accurate, this work often suffers from Holocaust myopia and, at times, intellectual bullying. Having spent so much time immersed in a topic so rich with moral relevance and historiographical detail, and one to which they often have a personal connection, it only seems logical that these scholars insist the Holocaust be covered with greater depth.

Nevertheless, their demands of teachers are unreasonable. Quite simply, most secondary school teachers do not have the time or the inclination to immerse themselves in topic of the Holocaust. Nor, unfortunately, do they have the pedagogical expertise necessary to engage in such a controversial and complex topic (or at least not to the extent that many scholars would like). A recent study reported that more than half of secondary social studies teachers do not even have a major or minor in history.[44] The proportion of elementary school teachers with a major in history is likely to be only a fraction of that. Compared to the average teacher, the innovative ones covered in this study were unusually qualified, informed, and committed. Scholars should be praising their ambition, not discouraging it with ivory tower critiques. The energy exerted trying to educate teachers about the Holocaust might be better spent trying to educate them about the discipline of history in the first place.

Overall, several recurring assumptions emerge from this critical literature. First, there is a scholarly consensus that any kind of Holocaust simulation automatically trivializes the event, simplifies the history, and leads to superficial emotional and cognitive understandings. Despite the enthusiastic endorsement of several teachers and the positive experiences of the game's designer (Rabbi Zwerin), scholars believe that any kind of reenactment is disrespectful to survivors and victims. In addition, low-level activities such as word searches, crossword puzzles, and multiple choice tests—which are all considered poor pedagogy to begin with—are deemed especially offensive when employed in the teaching of the Holocaust.

Second, there is a belief that the facts of the Holocaust must be covered fully, and that there is a positive correlation between moral development and the depth of historical understanding. The more students are immersed in the facts of the Holocaust, and the more time they have to consider its moral implications, the greater their ethical and cognitive growth will be. However, coverage seems to be defined horizontally across time, rather than vertically in terms of depth. In other words, if a teacher spent only three days on the event, is it assumed that s/he could not have done an adequate job, no matter what was taught or learned by students. In this way, coverage—defined narrowly in terms of topics and time—actually becomes an end in itself, not a means toward moral and cognitive growth.

The third assumption is that the more educated and informed the teacher is about the Holocaust, the greater the overall educational experience will be for students. Thus, the well-informed teacher is, as Wiesel suggested, the messenger, who merely delivers facts of the event and allows students to arrive at their own conclusions. Scholars seem to value knowledge of the event over pedagogical knowledge and classroom experience. They also overlook how teachers transform the content to deliver a number of implicit and explicit messages.

Fourth, there is an assumption that linking the Holocaust too directly to contemporary events or ecumenical prejudice will dilute the particulars of the event. Of course, the Holocaust is relevant to the present, but as historian Deborah Lipstadt and others have suggested, this connection should be implicitly, not explicitly, stated. Teachers need to explain the context of long-standing European anti-Semitism along with the Nazi theories of racial hierarchy. Such factors make the German anti-Semitism a distinct form of prejudice, not a generic form of discrimination.

Finally, the assumption underlying much of the criticism aimed at the published Holocaust curricula is that what appears on the printed page is what actually gets taught in the classroom. Of course, scholars critique these curricula because they can easily gain access to them, and because they are believed to be reflective of what is actually going on in the classroom. But it is presumptuous to assume that one can gauge how a teacher actually enacts a curriculum based only on its written contents, many of which are over three-hundred-pages.

These assumptions, along with numerous articles offering suggestions and resources for Holocaust educators, are based largely upon a priori theories and anecdotal evidence about how and why the Holocaust should be taught. These well-meaning scholars offer lists of "shoulds" and "oughts" mostly to the specialized audiences of research journals. However, what actually goes on in the classroom and how teachers employ these various Holocaust curricula and resources has, until recently, been a mystery. In the next chapter I will review what little empirical research exists on how educators teach the event and what students actually learn from it. This growing (and fairly recent) body of work challenges virtually every one of the scholarly assumptions listed in this chapter, and, in some cases, has proved them dead wrong.

Chapter 7

Out of the Discourse, Into the Classroom

As the Holocaust became more popular and pervasive in American culture, researchers became interested in how this came to be and why. The majority of the literature focused on one of two questions: first, could the Americans have done more to aid the European victims of the Holocaust during World War II? Second, how has the popularity of the Holocaust in America transformed memory of the event? The first question has been examined extensively. Historians like Arthur Morse, Henry Feingold, Monty Noam Penkower, and Martin Gilbert have argued that, in fact, the American administration could and should have done more during World War II to prevent the Nazi assault on the Jews. They point to the anti-Semitic U.S. immigration policies and the failure to heed Jewish pleas for aid and rescue.[1] David Wyman has even gone so far as to criticize the U.S. military for not directly attacking the rail lines to the Nazi death camps.[2] Similarly, Deborah Lipstadt has reviewed the American press coverage of the Jewish persecution during the war. She discovered that accounts of the Holocaust were often ignored, dismissed, or buried deep inside newspapers, because most editors and reporters literally found the stories "beyond belief."[3]

By considering the American response to the Holocaust these scholars have accessed the policies of the Roosevelt administration against what they perceive to be American ideals of pluralism, democracy, liberty, and equality. They ask why the American response to the persecution of the European Jews did not adhere to these ideals and suggest that Americans had a responsibility to do more than they did. This is one way in which American scholars have tried to reconcile the Holocaust with American culture. But the issue has been contested. Historians like Peter Novick and

Lucy Dawidowicz refute these criticisms by refocusing attention on America's role in defeating the Nazis.[4] They suggest that, at the time, the most effective way to end the Holocaust was to defeat the Germans as quickly as possible. This meant focusing all U.S. resources toward this goal, even at the expense of Jewish rescue missions. In addition, as I discussed in chapter one, many Jewish organizations were more focused on combating anti-Semitism and discrimination at home than they were on engineering rescue efforts.

The second theme, analyzing Holocaust memory in America, has been more pervasive in the literature over the past two decades. These scholars focus on how American cultural texts such as Holocaust memorials, literature, television shows, and films have transformed the meaning of the Holocaust to fit the needs of American society.[5] The process involves downplaying the Jewish elements of the Holocaust and concentrating on those that resonate with all Americans.

The exhibit at the United States Memorial Museum is often cited as the most conspicuous example of this process.[6] "There is a sense," writes Tim Cole, "in which bringing the Holocaust to Washington, DC, inevitably meant that the Holocaust would be Americanized."[7] Scholars, mostly sociologists, have offered their own interpretations of the Museum's exhibit and how it has managed to balance the Jewish and non-Jewish aspects. Edward Linenthal suggested that the Museum's exhibit offers four possible memories: a burdensome one, which challenges governments and individuals to prevent future genocides; a treacherous one, which evokes self-pity and incrimination; a murderous one, which suggests that mass murder is an unavoidable part of modern society; and a hopeful one, which inspires Americans to come together in mutual support.[8] James Young suggested, "the Holocaust memorial defines what it means to be American by graphically illustrating what it means not be American."[9] He explained:

> Such American icons of democracy will either be affirmed for the ways their ideals prevented similar events in America or, in the eyes of Native Americans, African Americans, and Japanese Americans, reviewed skeptically for the ways such ideals might have prevented, but did not, the persecution of these groups on American soil. Every visitor will bring a different experience to the museum, as well as a different kind of memory out of it.[10]

Young has been the most prolific American researcher on the modes of Holocaust memory in America and abroad. His particular focus has been international and national Holocaust memorials.[11] Having reviewed dozens of these memorials across the United States, he concluded that "liberty and pluralism" were "the central memorial motifs in both current and proposed museums."[12]

Like the historians assessing America's role in the Holocaust, these scholars have analyzed the Holocaust against a standard of American ideals. They have shown how Holocaust memory in America has been transformed to meet the perceived needs of contemporary American society. Many of these works often make passing reference to Holocaust curricula as another example of the "Americanization" process. As we shall see the concept of Americanization has also had an effect on how researchers have assessed the extent and effects of Holocaust education at the classroom level.

The substantial body of empirical research that has been conducted on Holocaust education ranges in it methodological approach and rigor. In the past decade, ethnographic studies of Holocaust education of individual classrooms have appeared in some of the top educational research journals including the *Harvard Educational Review*, *Curriculum Inquiry*, and *Teachers College Record*. For those readers not familiar with qualitative research, I will briefly describe its ontological and epistemological underpinnings.

Traditional quantitative research relies on sample size, methodological rigor, and the replicability of the data analysis to establish its validity. In this manner the results of quantitative research are meant to be generalizable across contexts, or at least representative of the sample under investigation. This approach dominated educational research for the first half of the twentieth century, and is probably still the most common form of research today.[13]

However, in the past thirty years, qualitative research has become more common among many educational researchers. Emulating the methodologies of cultural anthropologists, these researchers employ rich, thick description of specific classroom interactions in particular contexts. Often the authors openly reveal their own biases and backgrounds, and suggest how these may have affected their interpretations of the data. Meant to be suggestive rather than generalizable, ethnographic research focuses more on process of teaching than on particular inputs and outputs. As we shall see, Holocaust education researchers have used both quantitative and qualitative methods in their quest to discover to what degree the event is being taught, how it is being taught, and what students are actually learning from it.

Is the Holocaust being Taught?

In 1989, Joel Epstein of Olivet College surveyed Church-related institutions of higher education to determine to what degree his colleagues were teaching the Holocaust and how their students were responding to the instruction. Presumably, Church-related colleges and universities would be more likely than public institutions and secular private schools to address the Holocaust because of the Holocaust's obvious moral ramifications. These schools were far less likely to have departments of Judaic or Hebrew studies, so the courses would have been taught mostly by history professors. Having sent out questionnaires to 160 institutions, Epstein only received 37 back. Many of those who did not respond simply related that the Holocaust was not being taught at their institution. An additional 8 surveys were thrown out because the Holocaust was covered in a cursory manner as part of a general survey course. The respondents included those from all major Christian denominations. From his study we can get a general sense of how popular the Holocaust was as a voluntary topic among non-Jewish populations.

The instructors related how they had little to no institutional support for their Holocaust courses; they were taught as electives and were dependent upon the initiative of the individual professors. All of the respondents believed firmly in what they were doing and reported that their classes had a profound effect on their students. Some expressed that their Christian students often became "defensive," "angry," "perplexed," and "shocked" about the role of Christian anti-Semitism in the Holocaust, and had a difficult time reconciling their faith with the German perpetrators and bystanders. The majority of the respondents reported that most students "had been exposed to familiar stereotypes about Jews and encompassed them in their mind set in varying degrees." One professor reported that a student had suggested that the Jews were paying for deicide, and that many of his students were fascinated with Nazi helmets and gadgets.

One notable respondent to the survey was historian Christopher Browning (then at Pacific Lutheran University, now at University of North Carolina-Chapel Hill), who authored the classic Holocaust texts *Ordinary Men: Reserve Police Battalion 101 and the Final Solution in Poland* (1992) and *Fateful Months: Essays on the Emergence of the Final Solution* (1985). Browning reported that he too had taught one or two Christian fundamentalists who were convinced that God had punished the Jews. He was also less assured than his colleagues about the impact his interim class had on students, many of whom were "lemons" who "would take anything to get it over in a four week period."[14] While less than a whole-hearted endorsement of Holocaust education, Epstein's survey demonstrated that despite the draining and controversial nature of the content, there were many professors at Christian schools dedicated to teaching the event.

The similar study by Hubert Locke surveyed thirty-three American scholars in the field of public administration and public policy to inquire if and how they address the Holocaust in their courses. Only six responded that they did in "any substantial way." When asked if they considered the destruction of the European Jewry during the German Third Reich to have any major significance as a topic for teaching or research in their discipline, only nineteen responded affirmatively. From this data, Locke concluded that teaching the Holocaust was not a substantial concern in his field of public administration and policy.[15]

Stephen Haynes, who reported his findings in 1998, conducted a far more inclusive study of the Holocaust in higher education. Based on ninety responses from all types of Universities, Haynes confirmed that departments of history were bearing most of the burden of teaching the event. Most of these history departments offered the course as a regular elective offering. Unsurprisingly, the vast majority of professors reported using traditional methods of instruction such as lecture and discussion. However, like their high-school and middle-school counterparts, these professors regularly showed Holocaust films such as *Night and Fog* and *Shoah.* Elie Wiesel's *Night* was assigned by about 40 percent of the instructors, followed by Primo Levi's *Survival in Auschwitz* and Christopher Browning's *Ordinary Men* as the most commonly assigned texts.

Regarding the effect of the material on students, many respondents suggested that interest in affective elements of the event were beneath them. One responded "this is not a professional concern, in my opinion," and another wrote that the Holocaust, "is not a morality play, [the] the goal is critical thinking…for those interested in finding a permanent home place in the curriculum of academic institutions it must be treated as an academic subject." Haynes, a professor of religious studies at Rhodes College, was troubled by this neglect and asserted that Holocaust education in higher education should move beyond the mere transmission of content. He suggested that it should be directed toward "the formation of students' minds and characters."[16]

From these three studies we see that Holocaust education has made considerable gains at the level of higher education since the 1970s, although it is hardly a central concern of most institutions. The number of students being exposed to the Holocaust is likely higher when one takes into consideration the coverage it may get in surveys of Western civilization or in more broadly conceived courses on genocide. Nevertheless, the deep impact of the course on students—the kind reported by Elie Wiesel in his 1975 article (see chapter two)—did not seem to be occurring on any consistent basis. Many of the students taking these courses likely had a prior interest in the event, and/or had already learned about it. In fact, Holocaust education may have been a victim of its own success. The first university students to take Holocaust courses in 1970s were shocked and resentful of the neglect of the event and, therefore, could feel subversive by overcoming this "conspiracy of silence" or "null curriculum" by enrolling in classes on the topic. In contrast, today's students can no longer be enraged by such neglect; everyone had at least a basic outline of what happened in the Holocaust. Due to the pervasiveness of the Holocaust in popular culture, the event has lost its shock value and subversive appeal for students.

Despite its high profile in popular culture, the Holocaust has not become a major concern in areas outside the discipline of history. In 1998, Leona Kanter of Mercer College conducted a thorough study of college textbooks in sociology, political science, Western civilization, and American history, which were selected by her colleagues as the best in the field. She was hoping to get a sense of what kind of Holocaust knowledge students who never took a specific course on the topic would gain through a college education. Textbooks would provide the best way to determine this.

In sociology textbooks the number of lines devoted to the Holocaust ranged from zero to forty. She was appalled by the neglect of the topic in her discipline, which did the poorest job of addressing the implications of the event. "As a group," she reported, "these books failed to take advantage of the Holocaust as a way of understanding the link between social structure and social consciousness." The circumstances of Holocaust, she asserted, "are the very grounds for asking the classic question 'How is Society possible?' "[17]

As expected history texts offered far more coverage of the Nazi assault on the European Jews. American history texts included between 0 and 175 lines. One fascinating finding was the willingness of these American history texts to offer pointed critiques of the Americans' failure to intervene in the Holocaust. One text (Bernard et al. *Firsthand America,* 1993) reported:

> The American Government which made military detours to preserve the art of the Japanese city of Kyoto and the architecture of the German city of Rothenberg, sent no bombers to destroy the ovens at Dachau and Auschwitz or train trackage to them, although the administration had evidence available that amply suggested what the ovens were being used for.[18]

Another text (Lafaber, *The American Age,* 1989) wrote, "FDR's indifference to...the systematic annihilation of European Jewry emerges as the worse failure of his presidency." Many of the American history textbooks cited David Wyman's provocative 1984 book *The Abandonment of the Jews* and suggested it as further reading.

Unsurprisingly, Western civilization and world history textbooks devoted 10–295 lines to the event, although Kanter was surprised by "what has been left out of so many of these volumes." Ultimately, she concluded that "college students can complete survey courses in both world and American history with only a minimal confrontation with the Holocaust and then often in the form of a troubling photograph which may or may not gain their attention"[19] Kanter's study, along with the others, confirm that despite the rise of interdisciplinary Holocaust studies at certain institutions and the interdisciplinary units designed for middle- and high-school students, at the level of higher education the Holocaust was largely being left to the departments of history to cover.

If the Holocaust was being covered so poorly by textbooks and with such infrequency by college professors, perhaps state legislators were justified in mandating the event. But, if the objective of the mandates was to increase the coverage of the Holocaust in its schools, to what extent did this work? In 1998, the New Jersey Holocaust Commission, which had mandated the teaching of the event four years earlier, reported that 93 percent of it schools had assimilated the topics of Holocaust and genocide into their curriculum.[20] A 2003 follow-up study surveyed ninety-three school principals about their school's adherence to the New Jersey mandate. According to the surveyor, a little over half of the schools demonstrated an "acceptable" level of implementation. New Jersey teachers reported using resources and materials from Facing History and Ourselves Foundation, the United States Holocaust Memorial Museum (USHMM), and the New Jersey Commission on Holocaust Education, as well as using popular texts such as *Diary of Anne Frank* (Elie Wiesel's *Night* was not mentioned). Reading assigned texts and engaging in class discussions were the two most common instructional strategies. Most schools also reported going "beyond teaching only the historical facts to examine the role of prejudice, discrimination, stereotyping, and racism."[21]

A 2002 study by Jeffrey Ellison on the teaching of the Holocaust in Illinois, a state that mandated the teaching of the event, reported that teachers spent an average of eight hours of instructional time on the Holocaust. They mostly employed traditional methods such as lectures, discussions, and films, especially *Schindler's List*. Ellison revealed that "there was a tendency in Illinois high schools to subsume the topic of the Holocaust within the topic of tolerance and stereotyping" instead of the specifics of anti-Semitism.[22] Ellison discovered that the vast majority (88 percent) of Illinois teachers were teaching the Holocaust in their American history courses. Emphasis on Holocaust education tended to be higher in suburban areas with substantial Jewish populations of students, teachers, and administrators and in areas with high non-white ethnic student populations. Another significant finding of Ellison's study was that the single most important factor in determining whether a teacher would address the Holocaust in-depth was what he termed his/her "Holocaust profile"—the teacher's specific training, interest, and background. Accordingly, he recommended that the most effective way to increase instruction on the Holocaust was through teacher training and coursework, not state mandates.

A 2001 study of 254 teachers in Indiana confirmed Ellison's findings that most teachers learned about the Holocaust through "self-study," and they expressed a desire for additional training and preparation.[23] Indiana does not mandate the teaching of the event, but instead supports a Holocaust Committee. A 1999 investigation of the impact of the Arkansas Holocaust Education Committee's professional development conferences demonstrated that teacher training could have a substantial effect on classroom practice. Participants reported that after attending the conference, they were more likely to teach about anti-Semitism, the world's response to the Holocaust, Jewish resistance, Nazis rise to power, the role of bystanders, the totalitarian ideology of the Nazis, and other Nazi victims. They were also more likely to read and discuss first-person accounts of the Holocaust, and move from a single historical source to multiple sources. After the conference most participants increased substantially the number of class periods devoted to the topic.[24]

In 2004, Julie Patterson Mitchell conducted a study of seventeen award-winning Holocaust teachers in Tennessee, a state that does not mandate that teachers address the Holocaust and genocide, but instead supports a Holocaust Commission. The participants in the study, which included both social studies and language arts teachers, were interviewed about their instruction. They revealed that they focus their teaching "more on reasons of learning about humanity than on specific historical dates and similar information."[25] All seventeen of the teachers had participated in some kind of Holocaust-related professional development through Facing History and Ourselves Foundation, the USHMM, and/or the Tennessee Holocaust Commission.

The Tennessee teachers commonly employed both literature and film in their instruction. The most common books were Elie Wiesel's *Night*, Lois Lowry's *Number the Stars*, Jane Yolen's *The Devil's Arithmetic*, and *The Diary of Anne Frank*. Widely used films included *Schindler's List*, *The Courage to Care*, and *Not in Our Town*. Almost all of the award-winning teachers invited Holocaust liberators and survivors to personalize the history for students, and most of them also took students on field trips to Holocaust-related exhibits, museums, plays, and other community events. Although most of the participants used essays to assess the impact the lessons had on their students, they also employed more traditional assessments such as multiple-choice and matching tests. Overall the participants found the guest speakers to be the most effective pedagogical tool. As a result, the participants reported, "specific examples of seeing students act more positively toward others at school, or they revealed how the lessons had affected them through their writing assignments."[26]

Holocaust "coverage" is a nagging issue, and one that is not likely to go away. Of course, the Holocaust—or any historical topic for that manner—can and perhaps should be covered in greater depth. Unfortunately, curricular coverage is a zero sum game; advances in one topic lead to losses in another. In fact, the Arkansas teachers, when asked about the biggest barriers to implementing more content on the Holocaust, reported as the most common answers that "the curriculum is already too crowded" and "there is not enough time in the school year to cover this topic adequately."[27] Critiques of Holocaust neglect—past and present—fail to address this point. There is simply not room for everything, and so teachers, textbook publishers, and curriculum designers have to make difficult choices. At the level of higher education, taking a Holocaust course may mean not taking a course on the Civil Rights Movement or women's studies. The professors of these courses are not going to be convinced that the Holocaust is more significant and meaningful than their topic. All professors have myopia when it comes to their discipline or subject, not just those who study the Holocaust.

My own experience confirms the difficulty of having to make such curricular choices. I recently attended a workshop about the four-hundred-year commemoration of Jamestown, at which a panel of passionate, articulate Native Americans expressed their disgust about how their history and culture was being covered so superficially by Virginia textbooks. They volunteered to come out to schools as guest speakers. I considered asking one of them to come speak to my social studies methods class. The problem was I had already invited the scholar-in-residence from the National Slavery Museum, a teaching fellow from the USHMM, and a local lawyer who worked for Legal Aide to talk about poverty. Since I am primarily charged with instructing my preservice teachers how to teach social studies, I did not have any more room for the Native American speaker. Which topic should I have cut out to make room? Slavery? The Holocaust? Poverty? Each topic is equally deserving of coverage and representation.

The point is that these textbook and curricular critiques would be much more useful (but controversial) if they not only suggested what else needed to be covered, but what other topics should be curtailed to make room for the Holocaust. What should be cut out of the public school curriculum? Vietnam? Hiroshima? Rwanda? At colleges and universities, what should be cut out of the general education curriculum? Foreign language? World geography? American literature? All professors consider their topic highly significant and relevant, but, ultimately, students cannot take everything.

Coverage is important, and my narrative had demonstrated remarkable growth in Holocaust education in just thirty years from complete neglect to substantial inclusion. "I don't mean to say that we know as much as we need to about the subject," James Sheehan, professor of German history as Stanford University stated in 1995, "but I'm not persuaded that the Holocaust has been neglected."[28] At a certain point coverage will lead to overexposure, and then to Holocaust resentment and fatigue. Educators need to be careful that they do not reach that point. In addition, as many have pointed out, just because the Holocaust was being taught, does not mean it was being taught well.

How is the Holocaust being Taught?

We begin this section with what I consider the most provocative research ever done on Holocaust education, Simone Schweber's ethnographic study of a Holocaust simulation in a public school classroom. Recall that there has been almost universal consensus that simulations are pedagogically unsound, trivial, and disrespectful of survivors. Schweber entered the study with these same biases. She worried that a Holocaust simulation "necessarily collapses the important distinctions between the experiences of the Holocaust victims/survivors and present-day students, desensitizing students to those very real differences and allowing a 'shocking naivety' to masquerade as greater understanding of Holocaust atrocity." She feared that such an activity could possibly harm students psychologically and/or reinforce the image of the Jew as victims. Schweber admitted that she was mostly sure of her conclusions heading into the classroom and "was seeking empirical support for my claims." However, after observing the simulation and seeing its effect on students, her biases were overcome, and she was convinced that the activity was indeed pedagogically sound.[29]

The simulation was run by Ms. Bess in an elective course on World War II at a primarily African American school. Although the simulation was based on Zwerin's *Gestapo* game, it went far beyond his original design. Instead of the value markers (i.e., family, home, religion, life), which were to be risked on each turn in Zwerin's game, students were given an identity of an actual Jewish victim. The given identities were meant to correspond with the characteristics or experiences of the students. Ms. Bess also instructed her students to select two "cherished ones" to accompany them on their experience.

Ms. Bess designed the game, as she explained to her students, so they could "connect more personally and emotionally to the infamous World

World II event known as the Holocaust...Very simply, you and your cherished friends and family members will embrace the fate of Jews and learn what happened to us in each year we are studying."[30] The use of first person was a significant choice in creating this alternative world. Ms. Bess followed the outline of Zwerin's game by focusing only on the Jewish experience and by portraying the Holocaust not holistically as a finished, monolithic event, but rather as a series of events that unraveled incrementally over the course of the 1930s and 1940s. In this sense the definitional uniqueness of the Holocaust was preserved. Just as in Zwerin's game, students were viewing the event from the perspective of the Jewish victims, not from the perspective of the historian. The victims (and student-simulators) did not know what was coming next, and so they had to make decisions based on what they knew at the time. This "humbling" lesson was what Ms. Bess was trying to impart in her students.

The simulation spanned the entire semester, although the game was not played every day. Ms. Bess introduced the game on random days, in which a "roll call" had students scrambling to find their hidden identification cards under the threat of their character's death. If students were absent on a randomly selected Gestapo day, they risked the life of their character or that of their "cherished ones." In this manner, the game was also used as a form of behavior management, to increase student attentiveness and attendance. It was also intended to replicate the feelings of the victims' helplessness and the randomness of the Nazi persecution.

This odd juxtaposition portrayed by Ms. Bess throughout the simulation intrigued Schweber. "Ms. Bess's role consistently treaded the boundary between controller and comrade," Schweber explained, because she was "both 'tormentor' and 'savior,' 'torturer' and 'saint,' dualities with problematic moral implications."[31] Besides the ambivalent role of Ms. Bess, Schweber felt that the simulation had deemphasized the role of historical anti-Semitism, and that Ms. Bess had overlooked certain opportunities for moral growth. However, these paled in comparison to the accomplishments of the course. As Schweber explained, "Ms. Bess's simulation spelled out the possibility for students to deliberate very powerful moral dilemmas with a sense of real consequences."[32] In the end the researcher was "awed" by the accomplishments of the course. Students learned an impressive amount of information about the Holocaust and appreciated its moral dimensions.

Schweber's findings were controversial and directly contradicted the informed advice of the USHMM guidelines for teaching about the Holocaust. Her conclusions inspired a response from Miriam Ben-Peretz, an Israeli professor, who suggested that the strict rules imposed by Ms. Bess on the simulation prevented students from the open-ended inquiry that should underlie constructivist teaching. In a position very much in line with Samuel Totten, she suggested that students would have been better served by interacting "directly with a variety of sources and artifacts of the period and shape their own path." More significantly, echoing the many criticisms of *Facing History*, Ben-Peretz pointed out that the main objective of the curricular approach was "in the realm of values—to gain empathy and humanity in the face of that tragedy," not to understand its historical context. Understanding this context, Ben-Peretz asserted, should be the objective of Holocaust education.[33]

Schweber responded that her depiction of the simulation did not do justice to Ms. Bess who was far more pedagogically interactive and constructivist than she was portrayed. Schweber reiterated the main point of her ethnography—that simulations cannot simply be dismissed or endorsed on a priori theoretical grounds. She defended the simulation against the "fallacy" that only particular pedagogical arrangements are conducive to moral learning. All pedagogical-content systems, she asserted, "have the power to inform morally and transform informationally if done right, just as all have the power to fall flat, if executed badly."[34]

What emerged from this exchange was the significance of anti-Semitism to teaching the Holocaust. For Ben-Peretz, anti-Semitism was not only a context for the event, but it represented the primary explanation and an overall meaning of the event. As Schweber pointed out, Ben-Peretz's rationale for Holocaust education was mainly one of commemoration—an objective that may be appropriate for an Israeli students, but not for the African American public school students of Ms. Bess's class. Schweber also emphasized the importance of anti-Semitism, but endorsed it on disciplinary rather than commemorative grounds. For Schweber, understanding anti-Semitism was critical to developing historical empathy and disciplinary understanding—pedagogical objectives that were consistent with the latest social studies literature. Historical understanding was the criteria against which she was judging Ms. Bess's simulation, as well as the other curricula she studied.

Throughout the 1980s and 1990s, the *Facing History* curriculum had repeatedly been accused of ignoring anti-Semitism as the specific, not generic, form of discrimination surrounding the Holocaust. Unsurprisingly, Schweber's own ethnographic study of a Mr. Zee, a California teacher who taught a semester-long Holocaust elective course based on *Facing History,* confirmed these critiques. The population of Mr. Zee's school was economically, religiously, ethnically, and linguistically diverse, and his class reflected this. Throughout his course, Mr. Zee related personal, moralistic anecdotes designed to inspire his students to take action when they experienced injustice. Likewise, students freely shared their own personal experiences. The focus of the course was not on historical empathy, but on individual identity clarification. According to Schweber, Mr. Zee did not teach "students history on its own disciplinary terms but rather harnessed history to the heavy yoke of identity formation... Not a single historical primary source document appeared in [the] classroom... History was abnegated."[35] She found that while the class often talked about the Holocaust, they never really studied the actual facts of the event in any depth. Her interviews with students confirmed that they had learned little about the historical events of the period. For Schweber, it was not so much that Mr. Zee had not effectively achieved his progressive, affective objectives, but rather that these objectives were inherently flawed to begin with. The purpose of Holocaust education, according to Schweber, should be to understand the event on its own terms and in its own context, not to use the event for present purposes of identity formation.

In her book *Making Sense of the Holocaust,* Schweber compared her studies of four California teachers, including Ms. Bess and Mr. Zee. Her first case study depicted Mr. Jefferson, an instructor who raced to cram in as much information about the Holocaust as he could. Through such an approach, Schweber suggested, he effectively transmitted the factual knowledge of the event, but he "razed complex moral/historical terrain." For Mr. Jefferson the moral message of the Holocaust was transmitted as "dicta... stripped of [its] inherent richness and complexity."[36] As we saw earlier, Mr. Zee's emphasis on the affective elements of the Holocaust left his students ignorant of important historical facts. As a result, Schweber reflected, the Holocaust was "discussed as a symbol rather than understood as events," and his students completed the course with the "misguided impression that they had in fact learned a lot about the Holocaust."[37]

A similar outcome occurred in Mr. Dennis' class, whose Holocaust unit concentrated on a series of dramatic reenactments that Schweber described as "emotionally rich... but intellectually thinner than it might have been."[38] The most effective unit of the four was Ms. Bess's Holocaust simulation. Schweber also included discussions on the teachers' narrative emplotment, historical representation, and depiction of historical actors. For example, Mr. Zee ended his narrative with Holocaust survivors testifying how they overcame the event to live a full life. In this way the "story" ended on a redemptive note. In contrast, Ms. Bess's simulation game, which replicated the actual mortality rate, delivered a more accurate historical depiction of the event. Through Schweber's depiction of the explicit and implicit moral messages each teacher imparted, she paints a complicated picture of the allegedly "obvious" moral lessons of the Holocaust. Not only did the Holocaust "take radically different forms in different teachers hands," but, she suggested, it also delivered "widely divergent moral messages."[39] She concluded that mandated Holocaust education rests on faulty assumptions about the nature of Holocaust history and its moral dimensions.

In contrast to Schweber's response to the enactment of *Facing History* Melinda Fine published a pair of ethnographic studies of the *Facing History* in action that offered a far more positive evaluation. These articles presented a counterpoint to Schweber's evaluation, because Fine was much more comfortable with the progressive and affective goals of the popular *Facing History* curriculum. In fact, she openly endorsed them, especially for the racially diverse populations of urban settings.

In 1993, Fine reported the results of the *Facing History* on an inner city high school in Boston, Massachusetts, with an 80 percent minority population. The intellectual strength of *Facing History*, Fine explained, was "in its ability to help students feel personally connected to the subject matter [i.e., the Holocaust]," and to "connect students affective experience with a rigorous examination of a specific historical text."[40] The inner-city teachers appreciated the curriculum because they considered the Holocaust highly relevant to their students. Fine was initially dubious of the successes of the course reported by teachers, "given that the program's focus on twentieth-century European history is tangential to the cultural histories of minority students." But the teachers insisted, and ultimately demonstrated, "that the gap between present reality and historical events enables inner-city students to grapple with racism, prejudice, and violence more reality than if the course were immediately focused on them."[41]

To exemplify the successes of the curriculum, Fine focused most of the article on a set of class discussions related to accounts of anti-Semitism during the Holocaust. The discussions, Fine argued, "served as a window onto choices these students may make on a daily basis."[42] Throughout the course of the dialogue, students made connections between the events of the Holocaust and their own personal experiences with racism, prejudice, and retaliation.

Unlike what Schweber had done in her studies, Fine never assessed the students for their factual knowledge of the Holocaust. Instead she focused on the transformative effect the experience had on students. With this objective in mind, she offered *Facing History* high praise. She qualified her conclusions by insisting that for most students attitudinal and/or behavioral change occurred "slowly, unevenly, and at times too subtly to be readily discerned." But in the end, the *Facing History* created a climate "wherein students were able to recognize that there were a variety of viewpoints, identities, and interests in the world, all of which have some social grounding, and all of which must be understood if not necessarily accepted."[43] More significantly, Fine reported, students who were usually disaffected by the school curriculum were engaged in critical thinking—which was the ultimate goal of the course.

Fine's second study of *Facing History* was equally positive, but instead of focusing on the classroom environment, it offered an intriguing look at the interaction between teacher and student. Since the focus of curriculum was to challenge and complicate the thinking of students, controversy was not only addressed in class, but at times encouraged. Yet such exchanges could be intense, and teachers were often vexed by how to handle them. Fine's article focused on a heated exchange between an assertive student Abby and the teacher over the founding of the state of Israel. Abby was against "the way the country [Israel] was set-up," and insisted that "Elie Wiesel and people like that were involved, and millions of Palestinian people were massacred and forced to leave their homes, just like the Holocaust."[44] In response to Abby's assertive position, the teacher expressed unequivocal disapproval of the comment, but also reinforced that students should remain open-minded and respectful of difference. However, according to Fine, through the public (but, qualified) disapproval of Abby's position, the teacher used "her implicit authority... to undermine, rather than muzzle" Abby's position. "Given the power differential between teacher and student," Fine explained, "the critical debate between them is unevenly weighted."[45]

Uncomfortable with how the incident played out, the teacher in the study then invited a staff member from the Facing History and Ourselves Foundation to inform the debate. Abby again asserted her position and the guest speaker tried to convince her otherwise. "That's not quite true," the guest speaker insisted, "When the Israeli government came in, they didn't know what to do with the Arab land. The people that fled their home do end up living in camps, but that was the choice of the Jordanians, not the Israelis."[46] In the process of this discussion, Fine observed, certain voices were silenced and subordinated whereas others were empowered and privileged. Abby later revealed in an interview that she found the guest speaker close-minded and unwilling to acknowledge her view. Abby did soften her anti-Israeli position somewhat in light of the new information, but she continued to insist that the Israelis were oppressive and wrong.

From these exchanges, Fine concluded that neither the students nor the teacher and guest speaker "transcended their own political beliefs in interpreting and responding to political differences within the classroom." She insisted that this would be impossible to do, because political beliefs "are part of our internal make-up, as operative within these teachers as they are within our own interpretations of practice." In the end, Fine viewed the *Facing History* approach as promoting neither moral relativism nor indoctrination, but instead it offered a blueprint for forming multiple democratic communities inside the classroom and out. It fostered deliberation and critical thinking for the participants.[47]

Subsequent research by Simone Schweber confirmed Fine's findings about the role preexisting social, political, and religious beliefs have on the framing and teaching of the Holocaust. Schweber and Rebecca Irwin studied how the Holocaust was taught at a fundamentalist Christian school. The teacher, Ms. Barrett, focused her instruction around the text *The Hiding Place*, a personal memoir by Corrie ten Boom, a Christian who was put in a concentration camp for rescuing Jews. As Ms. Barrett explained, instead of focusing on the role of Christians in orchestrating or tolerating the persecution of German Jews, she used this text to teach students about persecution "that we, as Christians may someday face."[48] Unlike the many teachers cited in my narrative, Ms. Barrett had no ambitions of inspiring students to take action on behalf of social justice or genocide prevention. Instead, she aimed to strengthen their Christian identities, an objective she successfully achieved. As one student later explained, "I think [God], He was . . . kind of testing their faith, the Jews . . . because if they said that . . . they weren't Jews, then they wouldn't have been arrested and all that. So a lot of people were martyred there by saying they were Jews or saying they were Christians."[49]

The researchers argued that the Christian fundamentalism of Ms. Barrett and her students "shaped their historical understanding so thoroughly that other explanations for persecution during the Holocaust—such as biological racism or Church-based anti-Semitism, economic depression or modern functionalism—were 'occluded,' rendered invisible as possibilities."[50] The authors, who are both Jewish, found this depiction of the Holocaust troubling, especially in relation to the goals of multiculturalism, which aims to foster tolerance of others through celebrating difference. Instead, Bennett used the Holocaust to eclipse difference by making the Jews mere objects in a morality play about the righteousness of Christianity.

Schweber's study of the teaching of the Holocaust at an Orthodox Jewish school for girls confirmed that fundamentalist faith, whether it be Christian or Jewish, prevented open-ended inquiry into the event by providing predetermined answers and/or mystifying the secular events of history. Schweber observed Mrs. Glickman, the secular studies teacher who in addition to social studies covered English and science. Like Ms. Barrett, Mrs. Glickman taught the Holocaust to reaffirm her students' religious identities, not to convey specific moral lessons or to inspire social action. Over the course of the unit, Mrs. Glickman's students directed numerous questions at her, which she deflected with particularist claims about the incomprehensibility of the event. "Don't say 'because,'" Mrs. Glickman interjected at one point, "There was no because."[51] The question of "why,"

Schweber pointed out, so central to the disciplines of history and the social sciences, was considered superfluous in Ms. Glickman's classroom.

As a result of such instruction, Schweber explained, students treated their religious beliefs as ahistorical and made inaccurate assumptions about the historical actors they studied and judged. They also failed to collect evidence for their historical assertions, relating everything to God's awesomeness and incomprehensibility. When asked about why the Holocaust occurred, one student responded, "I think something had to happen because God made it happen... nobody could understand that for sure." Another student explained how she could not question God "because that's just the way it is. But you just have to have faith, even though I don't understand why such a bad thing could have ever happened and why *Hashem* [God] didn't stop it, I still have a lot of questions."[52] Mrs. Glickman's students learned about the facts of the event, but learned nothing about human nature, historical context, or about how to seek out the answers to their questions. As a result, students were not educated, but merely acculturated and indoctrinated further into their religious world.

In her most recent study, Schewber wrote a provocative ethnography of a third-grade teacher who taught the Holocaust to his students. Recall from the last chapter that Samuel Totten had directly attacked the notion of teaching the Holocaust to young children. This advice was not heeded by Mr. Kupnich, who viewed the event as a window into issues of social justice. "I've always tried to infuse my teaching with themes of importance of learning about intolerance and discrimination," he explained, "I think that's kind of the key point to any sort of education." Regarding why he needed to start teaching disturbing material at such a young age, he explained "if you start early enough,... then all of sudden it's not like this foreign concept that's just dropped on them in high school."[53]

Mr. Kupnich's unit consisted of reading numerous children's books about the Holocaust, which gradually intensified in severity and graphicness as the days passed. Students engaged in writing activities throughout the unit tracing their responses to what was happening. The unit culminated with life in the concentration camps and the discovery by Allies of the piles of dead bodies, images of which were shown to the third-grade students. The material had a profound effect on one particular student, Lila, the only Jewish child in the class. During and briefly after the unit, she entered a period of depression. "I get really sad," Lila explained "and I just... get all depressed and stuff, hearing about these people who, I mean, if I were born 50 years ago, this could have been me!"[54]

Overall Schweber was impressed by the ability of the students to distinguish between real and fictitious violence and, to an age-appropriate degree, be able to sympathize with the victims (and with Lila's grief). But, unlike the many teachers Schweber had observed, Mr. Kupnich had depicted the event accurately by covering the hasher aspects of the event adequately and ending the unit on a non-redemptive note. Despite the successes of the unit on a factual level, Schweber believed that the emotional effects of the Holocaust were too much for these young students to handle. She concluded that "the curricular creep [into the younger grades] ought to be curtailed *vis a vis* the Holocaust and that third graders, as group, are too young to learn about it in great detail." She seemed to be confirming Totten's claims about the pedagogical uniqueness of the event, for unlike the previous atrocity units Mr. Kupnich taught on slavery, Civil Rights, and Native Americans, Schweber, explained, "The Holocaust unit alone scared some of Mr. Kupnich's students" and "prompted nightmares."[55] She suggested that this could have been a result of the graphic imagery available for the Holocaust, which was missing from the other units. In fact, Keith Barton and Linda Levstik's research on young children has shown that historical images can have a profound effect on the ability of children to appreciate the past in more concrete terms, allowing them to conceptualize history in ways that verbal descriptions cannot.[56] In the end, Schweber asserted that just because third graders can handle the Holocaust does not mean that they should.

Taken together, this rich body of research demonstrates that the lessons of the Holocaust are far from obvious; they are divergent and can be easily shaped to meet the narrow needs of any community. Not only can the act of teaching take on a number of different forms and be directed toward a number of different objectives, but students themselves can walk away with a whole range of responses. These findings seem to undermine the effort to mandate Holocaust education, because learning about the event certainly can, but does not necessarily, lead to increased tolerance of others. In fact, state mandates were put in place before any real empirical research had ever been done on the effects of Holocaust education on children.

These studies further prove that pedagogical approaches to Holocaust education appear along a continuum with traditional, fact-based instruction on one end and progressive, interdisciplinary study on the other. The Jewish and Christian fundamentalist instructors tended to be more traditional in their teaching because they had specific messages to convey to their homogenous student populations. They were not trying to complicate or challenge students' thinking, but rather they were hoping to reinforce existing beliefs. Those who were teaching minority or heterogeneous populations tended to be more progressive in their orientation by universalizing the event and linking the German indifference and Jewish victimization to contemporary and personal events. They were hoping to complicate students' conclusions and break down their ethnic and religious stereotypes.

Since all of these disparate approaches and responses (besides the Orthodox Jewish one) can be considered part of the "Americanization" of the Holocaust, we can see that this term is not very useful as a pedagogical explanation. Aren't Christian fundamentalism and multiculturalism both products of American culture, even though they are opposites of one another? In the next section, we investigate what students really are leaning from the event.

What are Students Learning?

Once of the first rationales for teaching the Holocaust to children was that the event imparted moral and historical lessons, especially in regards to civic virtue. In fact, in his 1979 article in *Teachers College Record,* Henry Friedlander suggested that the Holocaust should be universalized and used specifically toward the goal of enhancing civic life.[57] But what exactly were the lessons of the Holocaust in relation to civics? In a 1996 study, Gregory Wegner asked students this very question. He analyzed the essays of two hundred eight graders, who responded to the prompt, "What lessons from the Holocaust are there for my generation today?"

Wegner's findings were consistent with much of the anecdotal evidence that appeared in the literature about the effects of Holocaust education and its lessons. The vast majority of responses addressed moral prescriptions about what students should not be doing. The most popular response (82 percent) was to not allow the Holocaust to happen again. "We must never close our eyes to the killings of many innocent victims and martyrs," one student wrote. The second most common lesson (64 percent) was not to dehumanize others. "Hitler and the SS were driven by hatred against the people who were different from their idea of perfection," another student explained. The third most common response (60 percent) was not to be a bystander. One student asserted, "People must be willing to speak out against things they believe are wrong." Other common responses included to not discriminate against individuals or groups, and to not follow political leaders blindly. There were two additional findings of interest. First, twelve of the respondents, who were apparently not used to having to synthesize and apply historical information, simply listed facts about the Holocaust in their essay without any reference to lessons learned. Second, the majority of responses (66 percent) perceived the Holocaust as solely a Jewish event, with no mention of other victims.[58]

It is difficult to determine from these responses whether students really internalized these lessons or whether they were just reporting what they thought teacher wanted to hear. The study points to the difficulty of identifying with any clarity what the objective of Holocaust education should be. While rationales like "increasing tolerance" and "enhancing civic virtue" are admirable, they are not necessarily measurable. What exactly does increased tolerance look like and how would you assess it, especially in terms of it transferability to situations outside of school? There have been several studies that have attempted to quantify and assess the results of Holocaust education more specifically against that of control groups.

A 1981 study of *Facing History* by an outside reviewer employed a number of assessment instruments including the Kohlberg Standard Scoring Interview for moral reasoning, Selman's interpersonal understanding interview, and the Loevinger Sentence Completion Test for ego development. First, the researcher discovered that eight-graders do not like to take tests, which forced him to throw out much of the data, including the results of the Kohlberg moral assessment. However, with the reliable data he did collect, he discovered that students who were taught the unit increased significantly in their interpersonal understanding as well as factual knowledge about the event.[59]

A 1982 study of eight-grade student journals, which used Piaget's stages of epistemological development, reported that *Facing History* increased students' in-stage cognitive development and reflective thinking. A 1994 study, which listed Margot Stern Strom as one of the authors, was aimed specifically at disproving an accusation that learning about the Holocaust was psychologically damaging to students. Using Rest's Defining Issues Test, the authors reported that the curriculum significantly increased eight-grade students' moral reasoning, and did not affect their psychological well-being. A 2001 study found that students of *Facing History* increased relationship maturity and decreased racist attitudes and self-reported fighting behavior more than the control group. However, there were only minimal gains in moral reasoning. Despite using large numbers of subjects, these studies were all conducted in the Boston area in schools with direct connections to Facing History and Ourselves Foundation. With the exemption of Schweber's enthographic study of the California teacher, there are no studies of the effectiveness of Facing History outside its direct sphere of influence, despite the fact that thousands of teachers are using the curriculum across the country.[60]

A 2002 study of *Facing History* by Veronica Boix-Mansilla had much clearer and humbler objectives than increasing tolerance. In addition her study was much more focused and grounded in cognitive research. Boix-Mansilla studied the effects of the *Facing History* curriculum on the historical understanding of twenty-five eight-graders (also in Boston). Although *Facing History* was originally written from more of a progressive, behaviorist perspective, the activity under investigation in Boix-Mansilla's study was more disciplinary in nature. Unlike Wegner, Boix-Mansilla was less concerned with the content of student answers than she was with process of constructing them. She focused on history as process rather than a body of evidence. This demonstrates how the changing curricular milieu (see chapter six) not only affected the construction of the curriculum but also how it is enacted and assessed.

The classes in this study spent between six and ten weeks studying the Holocaust and then watched a documentary video on the genocide in Rwanda. In response to this content, the students were asked to write an essay hypothesizing reasons for the killings in Rwanda. In doing so, they were instructed to consider the similarities and differences between the Holocaust and Rwanda in light of the testimony of a Rwandan Tutsi woman. The written responses were evaluated for evidence of historical modes of thinking.

Boix-Mansilla found that most students were able to distinguish between the different historical conditions surrounding the two genocides, the incremental steps involved in both, and actions and dilemmas faced by individual rescuers and victims. However, with few exceptions, "students failed to recognize the constructed nature of the very accounts on which they were grounding their hypotheses and interpretations about contemporary Rwanda." That is, according to Boix-Mansilla, they failed to appreciate that "narratives are humanly constructed, that they embody particular world views, that they are written with a contemporary audience in mind, that that they seek to be faithful to the life of the past."[61]

For example, a student that failed to think historically outlined how he would research the Rwandan genocide in the following way: "I would get all the info I could about the genocide and research and memorize it. Then after memorizing it, I would visit as many survivors as possible and get their side of the story. After this I would compare what the people said happened with what really happened." On the other hand, a more advanced student, who could think historically, wrote the following: "I would try to look at the problem for all sides and would read up on it by researching in libraries... I would never be able to have total knowledge on the subject without being there. Even then I wouldn't completely know because I couldn't know all the sides to the story." The first epistemologically naïve student viewed history as having happened one way and suggested that the job of the historian was to sort through those testimonies in order to dismiss the truthful ones from the false. More accurately, however, the second student understood that there was no "true" history outside the one that could be constructed intertextually from the multiple testimonies.[62] The second student could think more like a historian does, which was the main objective of the activity.

Unlike many of the other studies mentioned here, which either viewed content knowledge of the Holocaust as an end in itself, or used the Holocaust for personal transformation, Boix-Mansilla' study viewed the Holocaust as a means of acquiring historical skills and understandings. An in-depth study of the Holocaust, she concluded, "challenged student to build multicausal explanations of the events, ... to confront conflicting narratives; to consider various historical actors' points of view; to examine evidence; and to attend to continuities and changes over time." She hoped that these skills would be transferable to other contemporary situations. History, she suggested, "prepares students to examine the present by understating the myths, oversimplifications, and distortions embodied in popular views of the past."[63] Therefore, history instruction—and Holocaust education as subset of that—should be aimed specifically at developing the cognitive tools to think more like a historian.

Judging Holocaust education as a process instead of the contents of a curriculum and defining learning as cognitive growth instead of the accumulation of facts have both been significant characteristics of recent research. Both Fine and Schweber's ethnographies addressed the important issue of how the Holocaust takes on different meanings based on the political and cultural backgrounds of the teacher and students. In cognitive terms, students and teachers have preexisting cognitive structures that process the facts and emotions of the Holocaust by linking them to preexisting narrative schemes. Thus, Holocaust education is a dialectical process between the individual and content, not the direct transmission of specific facts and morals. A fascinating and disturbing study by Karen Spector confirmed this very point.

Spector studied the results of Holocaust units based primarily on the reading of Elie Weisel's *Night* in three different public school English classes in a Midwestern region. Specifically, she examined how students' employed their preexisting Christian cosmologies to "emplot"—that is, place the Holocaust in a particular narrative structure—the dissonant information from Wiesel's text in ways that were both comfortable and predictable to them. Students were specifically asked about the scene in Wiesel's *Night* in which Elie questioned his faith in God as he viewed a young boy hanging from a gallows at Auschwitz.

The majority of these public school students (69 percent) used religious narratives of supernatural intervention. In this manner, they imposed order on the events of the Holocaust in three ways. First, students suggested that both God and Satan were historical actors engaged in a struggle of good and evil; Hitler embodied Satan, and God ultimately saved the Jews by stopping the extermination before it achieved it ultimate goal. "It has been suggested that Satan killed all those thousands of Jews through

Hitler," one Christian student explained, "Evil can't penetrate things that are not evil, so I think Hitler had something to do with it."[64] Second, students expressed how there were certain ways in which individuals should act, and if they do not, there may be supernatural consequences. "God... let it happen for a reason," another student explained, "he was there the whole time, but since people were questioning Him and losing faith, He wasn't doing nothing about it."[65] Third, students employed narratives of "the cross" in relation to Jesus—that He was either suffering at Auschwitz alongside the Jews, or that the Jews were being punished for their rejection of Him. Overall these narratives, according to Spector, blamed the victims for their suffering, moved cause-and-effect from the natural to the metaphysical realm, and justified God's lack of intervention in the event.

Spector's provocative findings confirmed Schweber's—that the moral lessons of the Holocaust are not at all obvious or convergent. Even when using a text like *Night*, which is both popular and endorsed by numerous scholars, students extract a wide range of lessons from the facts. "Redemption from 'the gallows' and condemnation from 'the cross,'" Spector pointed out, "both position Jews and the Holocaust in ways that do not conform to the project of becoming more tolerant."[66] Locating the Christ figure in Auschwitz, besides de-Judaizing the story, removed the event from physical reality and historical context. In the end, not only did teaching about the Holocaust fail to transform certain students, but it appeared to have actually reaffirmed their existing belief systems.

The State of Holocaust Education in America

Taken as a whole this body of research supports a number of conclusions about the current state of Holocaust education. First, although state mandates, the content of textbooks and the amount of teacher training all impact the extent to which the Holocaust will be taught, the most influential factor is what Ellison called the teacher's "Holocaust profile." In other words, teachers with a personal interest in the event are more likely to do research, track down resources, and enthusiastically attend Holocaust workshops than those have no intrinsic interest. There is not a single example of a teacher who became interested in the topic as result of a mandate. In fact, all the exemplary teachers covered in these studies had been teaching the event for years before the mandates were implemented. In addition, teachers with Holocaust profiles tend to agree with the suggestions put forth by leading Holocaust educational organizations such the USHMM and Facing History and Ourselves Foundation, mainly because they have participated in professional development or used materials from these very organizations.[67]

The second significant finding is the extent of the popularity and influence of the Facing History and Ourselves Foundation, which along with the USHMM are the two most pervasive forces in Holocaust education. This is interesting because, as I have argued throughout my narrative, the two organizations offer conflicting approaches to teaching the event. Facing History and Ourselves supports an agenda of social justice and social activism, and frames the Holocaust in a progressive manner that will further these goals. On the other hand, the USHMM places more emphasis on the particularities of the Holocaust and seeks to impart an appreciation of the historical and definitional uniqueness of the event by focusing on anti-Semitism and engaging directly with survivor testimonies (although the museum also provides materials on other victims). Both organizations receive funding from the federal government, and both support grassroots efforts to improve and spread Holocaust education. Teachers do not seem to be cognizant of these subtle differences between the two organizations, and many ambitious teachers have attended professional developments sessions for both.

The third significant finding is that Holocaust education seems to be most popular in two areas—suburban districts with substantial Jewish populations and urban areas with high non-white ethnic populations. The Holocaust is relevant to both these areas, but for different reasons. Obviously, areas with high Jewish populations are more likely to have teachers with Holocaust profiles, who have greater knowledge and interest in the event. Teachers of inner city minority students teach the event as an indirect way to deal with the prejudice and discrimination that their students experience on a daily basis. The context of the school, to a large degree, seems to dictate the pedagogical approach. Teachers with large Jewish populations seem to emphasize the particular aspects of the event. On the other hand, teachers with either high minority ethnic populations or with ambitious goals of furthering social justice seem to employ a more progressive approach.

Finally, the fourth significant finding is teachers at the higher, secondary, and middle-school levels all seem to be relying heavily on three resources: Elie Wiesel's *Night*, *The Diary Anne Frank*, and *Schindler's List*. The former two have been popular with teachers for decades. In fact, I read both books in my eighth-grade English class (my Jewish teacher had a significant Holocaust profile). Wiesel's *Night* was selected for the Oprah Winfrey book club in 2006. Spielberg's movie has more recently become the "the film" on the Holocaust, at least in American schools. So, despite the numerous books, films, and memoirs published on the event and the extensive efforts of the USHMM and the Facing History and Ourselves Foundation, the major content and lessons of the Holocaust may simply be boiled down to these three major sources. Much more research is needed on what students and teachers actually learn from these texts.

As we revisit the a priori assumptions from last chapter, we see that few of them have held up to empirical study. First, Simone Schweber's study of Ms. Bess has challenged, if not disproved, the assertion that Holocaust simulations inherently trivialize the event. Although Schweber does not endorse simulations outright, she suggested that, in the right hands, they can be a powerful and effective pedagogical tool. Second, Schweber and Spector's research has demonstrated that merely covering the event, even in great depth, will not necessarily lead to greater tolerance, historical understanding, or civic virtue. The obvious lessons of the Holocaust are apparently not so obvious.

Third, this research has shown that even when teachers have Holocaust profiles and are very knowledgeable about the event, students do not necessarily benefit from this content. Teachers can employ their knowledge toward a number of different pedagogical objectives and ends, which do not necessarily include the transmission of more information. However, professional development does seem to inspire teachers to cover more topics and spend more class time on the event. Fourth, the assumption that linking the Holocaust too directly to ecumenical prejudice will dilute the particulars of the event does seem to be true. Students under the more progressively oriented teachers learned little about historical anti-Semitism, and often made casual comparisons to other events too easily for the tastes of these researchers. Finally, the assumption that what is published in a Holocaust curriculum is what actually gets taught in the classroom is too simplistic. As we have seen, *Facing History* had been directed toward both progressive and disciplinary ends.

This research has shown the great diversity and complexity of teaching a historical event as culturally important and morally laden as the Holocaust. Drawing upon the history of how the event had been taught as well as more recent cognitive theory, in the next chapter I attempt to bring some direction and order to Holocaust education. I suggest that the objectives of Holocaust education have been too ambitious, and that teaching about the event needs to be placed in the greater context of the goals of secondary education.

Chapter 8

Teaching the Holocaust and the Aims of Secondary Education

Teaching the Holocaust in American schools has been so controversial for a number of cultural reasons. First, secondary schools are one of the few public spaces where the multiple ethnicities, religions, and races are forced to achieve some kind of compromise. At the elementary level, there is general consensus that schools should focus on reading, writing, and arithmetic. However, at the secondary level, with a curriculum that includes the transmission of cultural content, things get much more complex. A textbook and curriculum cannot possibly cover everything, and so certain difficult choices must be made that inevitably involve multiple perspectives coming together. While some have suggested that it is morally wrong to compromise a topic such as the Holocaust, it simply cannot be covered in the depth it deserves without bumping out other topics.

Second, the media attention surrounding the Holocaust had brought a disproportionate amount of attention to minority positions. The school curriculum is an easy target for small, disgruntled groups to draw attention to their causes. As we have seen the press has reveled in the controversies surrounding Holocaust and genocide education. To be fair, reporters have represented the different elements of the debate fairly accurately, and—especially in the many political attacks on these curricula in this study—have even represented the voices of teachers.[1]

Third, perhaps more than any other event, the Holocaust inevitably leads to discussions of morals and values. As we have seen, if teachers decide to link the event to current examples of prejudice, they risk upsetting scholars who view such comparisons as ahistorical, or upsetting parents who believe that discussions of values should be left to families. On the other hand, if teachers choose to focus on the historical particularities of the event, they must address Christian anti-Semitism, a topic that has proved to be equally controversial. Either way, confronting and explaining the complexity of evil is a messy business. Comprehensive secondary schools were specifically designed to bring different social, ethnic, and academic groups together under one roof. As we saw in the previous chapter, several researchers have lauded Holocaust education, specifically *Facing History*, for fulfilling the promise of the secondary school by addressing issues of values and politics in a meaningful way.

Finally, many proponents of Holocaust education realize that only about half of Americans will ever attend college, and most of them will only take a few required classes in arts and sciences. Even though Holocaust courses are being offered in greater numbers on college campuses, the research confirms that these classes are always voluntary electives and mostly reach those students who have a preexisting interest in the event. So only a fraction of students learn much about the Holocaust at the University level. More Americans learn about the event from movies such as *Schindler's List, The Pianist,* and *Life is Beautiful.* Although scholars debate the accuracy of these films, they are first and foremost creative endeavors meant to make money for their investors. Scholars and proponents of Holocaust education have little to no control over the content of these films. Instead they must either choose to endorse or critique them. On the other hand, the United States Holocaust Memorial Museum has reached millions of Americans, and there are numerous other museums across the country. These museums have had a profound effect on many teachers who have drawn inspiration from their visits. Nevertheless, these museums reach only a fraction of Americans, and exactly what they learn from these exhibits is a mystery.

That leaves American secondary schools as the most efficient and pervasive way to disseminate awareness about the Holocaust. In fact, a 1992 study by the American Jewish Committee reported that 36 percent of adults and 59 percent of students listed school as their main source of Holocaust awareness.[2] But inclusion in the school curriculum comes at cost—what many have called the "Americanization" of the Holocaust. I prefer to think of it as the "educationalization" of the Holocaust, because teachers have transformed the event, not for cultural reasons, but rather for pedagogical ones—not to appease political interest groups, but rather to reach, engage, and transform their students. That is, the cultural reasons for the controversial nature of the Holocaust listed earlier are significant, but they do tell the whole story. As I have been arguing throughout this narrative, the debate over Holocaust education is more accurately conceptualized as a debate over the teaching of history.[3]

In this chapter, I will further develop the argument I have been pursuing throughout—that the debate over Holocaust education should be conceived of a continuation of the debate over the objectives and methods of teaching history to secondary students. I will then resolve this debate by suggesting how cognitive theory can provide us with a consistent, overall rationale for Holocaust education and for secondary education more broadly conceived. Such a rationale requires the resurrection of Lawrence Kohlberg's cognitive developmentalism—the very theory that helped to launch Holocaust education in the first place. I argue that Holocaust education needs to direct its moral and cognitive objectives toward more specific, identifiable, and obtainable goals.

Three Approaches to Holocaust Education

Holocaust education did not emerge in a vacuum. The affective revolution was an immediate context for the initiation of the movement, but its long-term trajectory must be viewed in light of preexisting curricular conflicts. Throughout the narrative I have referred to three primary approaches to teaching history: the traditional, the progressive, and the disciplinary. I do not mean to reify these approaches, nor to imply that practitioners and theorists used these names in any consistent sense. In fact, any curriculum likely contains elements of all three approaches. However, most curricula emphasize one above the others, especially in term of its educational objectives.

The traditional approach to teaching history was and is the most instinctual and common form of instruction. It involves students learning the facts of history chronologically through teacher lecture and textbook reading. Students are then tested for acquisition of this knowledge through "fact recall." Unlike "the mental discipline" approach, which involves the recitation of memorized text, the traditional approach often included multiple choice and essay questions, inspiring students to "think" about their memorized facts. Nonetheless, to know history for the traditionalist is to be able to recall, recognize, and organize the facts of history. The founders of the social studies largely constructed their curricular approach in direct opposition to the irrelevant, outdated traditional one. Despite the slow, but qualified success of the social studies movement, traditionalists continued to defend their approach throughout the century.[4]

By as early as the 1940s, traditionalists were arguing that encroachment of the social sciences into history instruction had eroded the factual knowledge of American citizens. Instead of the facts, students were being taught nebulous skills that enabled them to adjust to life more efficiently. In 1938, John Dewey identified this controversy as one between "traditional and progressive education," the former defined as "development from without" and the latter as "formulation from within." In *Experience and Education* Dewey attempted to reconcile this unnatural divide, insisting that students needed to learn the facts, but the facts also needed to be connected to the students' intrinsic concerns.[5] The traditionalists exploited this perceived division to separate themselves from the progressives, who they considered anti-intellectual and ineffective.

Mainstream historians, many of whom felt disenfranchised from the social studies curriculum, tended to lean toward the traditionalist position. They led a wave of criticism that erupted in the 1940s and 1950s. In a number of *New York Times* articles, historians cited survey results exposing the alleged ignorance of American youth about the facts of American history. They attacked the popular, problem-centered textbooks of Harold Rugg and blamed the progressive reformers for watering down the curriculum.[6] The most publicized and effective critique of progressives was historian Arthur Bestor's *Educational Wastelands: The Retreat from Learning in Our Public Schools.* Bestor attacked the progressives whom he accused of becoming "confused about the purposes of education,... by setting forth purposes... so trivial as to forfeit the respect of thoughtful men." He supported a liberal education that transmits the common knowledge of the past to all citizens through "disciplined intellectual training." He mocked those who "teach children" instead of teaching history.[7]

The most significant proponent of traditionalism in recent years has been E.D. Hirsch. His 1983 article "Cultural Literacy" provided a rationale for why all Americans needed a base of common "translinguistic" knowledge to be successful in society. First, Americans needed a common set of referents to define themselves as a culture. Just as students learning French had to study the French culture and lifestyle to appreciate and comprehend the language, Americans also needed to define and learn their own culture in order to be able to communicate with one another—an idea, he pointed out, that the Founding Fathers understood. Second and more significantly, his research on the reading skills of college students demonstrated that those who had greater background knowledge could better comprehend text. Even when two students had the same reading ability and read the same text, the one with greater background knowledge would have better comprehension. "Words are not purely formal counters of language," he explained, "they represent larger underlying domains of content."[8] In other words, he argued, reading was not simply a formal skill, but to a large degree was dependent upon a store of concrete background knowledge. Hirsch's research was influential because this was the first time that traditionalism could be defended on empirical rather than merely theoretical grounds. It justified what many had been saying for decades. He later expanded this idea in a book with the same title, which included a list of referents that cultural literate citizens needed to know.[9]

In contrast to the fact-driven traditional approach, the disciplinary approach to teaching history encouraged students to "do history" and "be historians." Fred Morrow Fling and Mary Shelton Barnes developed the disciplinary approach as early as the 1890s. Barnes advocated for a positivistic approach to writing history, but Fling was influenced by the new historians such as James Harvey Robinson and offered a more pragmatic approach. Fling's goal was to teach students "the process by which we attain to historical truth—in other words to teach historical method." Through this, Fling argued, students would learn that "knowledge grows and certainty is attained though question and answer," and they would appreciate "how difficult it is to arrive at certainty." Ultimately, he hoped they would be able to distinguish "between good and bad, scientific and popular secondary [historical] works."[10] Very few teachers adopted Fling's approach and those that did soon abandoned it in favor of other more modern approaches. It wouldn't be until the early 1960s that the disciplinary approach again entered the curriculum. The impetus for this movement would be the work of influential Harvard psychologist Jerome Bruner.

In *The Process of Education* Bruner presented his argument "that any subject can be taught effectively in some intellectually honest form to any child at any stage of development."[11] By subject Bruner did not mean the facts and details of a particular body of content. Instead he meant the subject's underlying "structure"—the discipline's central concepts, generalizations, and mode of inquiry used by scholars in the field. He suggested that students would learn more and retain the information longer, if they arrived at the facts themselves through carefully controlled experiments. The transmission of the basic structures of the field was more important than "covering" a superficial survey of the information in a field. For history this translated into gaining an understanding into how historical knowledge is constructed. Students could sift through strategically designed packages of primary sources to not only learn the historical content, but also acquire an understanding of the "structure" of the historian's discipline.[12] By the mid-1960s historians at colleges such as Amherst and Carnegie-Mellon designed their own primary-source-based projects and piloted them in local schools.[13] The most significant proponent of the disciplinary approach in recent years has been Howard Gardner. Gardner is mostly known for his misunderstood theory of multiple intelligences (MI), which many educators have mistaken for a theory of learning. MI theory is actually an alternative explanation for how society defines and assesses intelligence. To understand Gardner's theory of how people learn, one must turn to two lesser know texts—the *Unschooled Mind* and the *Disciplined Mind*.[14] Drawing on the work of Dewey, Gardner argued that multiple intelligences work together in specific domains to employ socially constructed, abstract sets of symbolic systems. The symbolic system for math is numbers, for literacy it is letters, for dance it is movement, for music it is notes, and so on. Learning is defined as the process of appreciating and internalizing these discipline-specific symbolic systems to help students overcome the naïve conceptions of the world they initially construct. "The disciplines," he asserted, "represent our best efforts to think systematically about the world . . . each discipline had developed its own means, its own 'moves' for making sense of initial data."[15] In history, the naïve conception (one that many adults still hold) is that a single, objective past exists and that historians look at primary sources to access and report the past directly.

This relationship to the discipline is significant because in opposition to Piaget and Kohlberg, Gardner argued that intellectual development is domain specific, not a series of overall epistemological shifts. Throughout the *Disciplined Mind*, he used the example of the teaching of the Holocaust and even suggests employing some role-playing and simulation activities. Students could "re-create dramatic scenes from the Holocaust, such as the defense of the Warsaw Ghetto, or debates within families about what to do when they were about to be separated from one another, or reactions among members of a unit when a solitary soldier refused to join in the massacre of Jews."[16] Gardner suggested that such moral role-playing, by putting students in the mentality of historical figures, would allow them to think more like a historian.

Finally, the progressive approach to the teaching of history has the strongest historic roots in the curriculum. Overall, this approach emphasized the continuity, commonality, and underlying societal truths of history and downplayed progression, causality, and historical context. Perhaps Dewey defined it best when he said that history should be approached as an "indirect sociology." History supplied the factual raw material for systematic analysis from which students could draw generalizations. Students were then encouraged to apply the historical facts and lessons to contemporary problems. The NEA's 1916 Committee on the Social Studies launched this approach, which was later implemented by the designers of many of the first Holocaust curricula.

In the eyes of some, the progressive, social studies approach to the Holocaust inevitably distorts history by downplaying its historical particularities and emphasizing certain features that may be deemed interesting and relevant to students. On the other hand, a traditional, straightforward account of the Holocaust that respects the definitional and historical uniqueness of the Jewish experience may not resonate with or even interest students at all (i.e., history for history's sake). Through the topic of the Holocaust, one can see an inherent tension between the traditional and progressive approaches to teaching history—a tension that did not originate with the teaching of the Holocaust, but actually stretched back to the 1890s. Educational psychologist Samuel Wineburg identified this tension in studying history as that between the "the familiar and the strange, between feelings of proximity and feelings of distance in relation to people we seek to understand."[17] Wineburg had been a proponent of the disciplinary approach, which in many ways can be viewed as a compromise between the traditional and progressive, because the disciplinary approach is both student-centered and inquiry-based, but still retains the rigor of "real" history.

The curricula covered in this study represented the different curricular approaches to teaching history available to teachers in the 1970s, including the traditional, the progressive, and the disciplinary. The particularist approach was suggested by certain Jewish scholars, but ultimately rejected by public school teachers. The controversial nature of the Holocaust as a topic has demonstrated that each of these orientations has certain benefits and weaknesses that are inherent in each particular approach. Ultimately, these weaknesses are unavoidable. The traditional approach, characterized by Albert Post's *The Holocaust: A Case Study in Genocide,* and Carol Danks and Leatrice Rabinsky's *The Holocaust: Prejudice Unleashed* simply related the facts of the Holocaust chronologically. The teacher dictated the moral message of the event and chose the meta-narrative in which to place it. In Post's case, he chose a redemptive narrative by ending his unit with the founding of Israel. Post's traditional approach emphasized the historical context of the Holocaust—including Jewish life before the catastrophe and the roots of anti-Semitism—but failed to make strong connections with contemporary events or to make the event resonate with non-Jewish students. In Danks and Rabinsky's case, they took a more somber approach by ending their curriculum with a lesson on survivors. In both cases, emphasis was on the past as a collection of evidence and facts, not on the present.

The progressive approach, employed by Richard Flaim and Edwin Reynolds in *The Holocaust and Genocide: A Search for Conscience* and by Margot Stern Strom and Williams *Parsons in Facing History And Ourselves,* organized the historical events of the Holocaust around an inquiry into human behavior. This approach engaged students with relevant material and enabled them to apply the lessons of history to contemporary events. Since answers about moral responsibility were often left unresolved, this approach avoided the moral indoctrination of the traditional approach. But, the progressive approach de-emphasized (or even distorted) the historical particularities of the German anti-Semitism by depicting it as an example of ecumenical prejudice. Such an interpretation, historians pointed out, was not only irresponsible, but historically inaccurate.

The more affectively oriented progressive approaches employed by Rabbi Zwerin in *Gestapo* and Roselle Chartock in *Society on Trial* reveled in the moral ambiguity of the historical actors under investigation as a way to have students clarify their own values. This approach also avoided moral indoctrination, and it questioned students' notions of historical inevitability. But like the more behavioral approaches, the affective approaches left students with little understanding of historical context and continuity. In addition, critics questioned whether changing students' attitudes through emotional shock equated to "liberal" indoctrination. Thus, curriculum designers abandoned the affective approach by the end of the 1970s in favor of the more "scientific" inquiry-into-human-behavior approach. But even this orientation had its shortcomings, which eventually led Holocaust curriculum designers to move in a more disciplinary direction.

Each of the curricula covered in this study did include the use of some primary source material. Thus, the disciplinary approach was employed to some degree. But until the United States Holocaust Memorial Museum (USHMM) Resource Book, the disciplinary approach was not used as the primary mode of investigation for three reasons. First, one of the main objectives of teaching the Holocaust was to overcome the conspiracy of silence surrounding the event. Teachers were appalled that a historical event so important was being systematically neglected in the curriculum. They introduced the Holocaust to relate the facts that textbooks failed to cover. The disciplinary approach would not have been the most efficient way to relate the facts of the event, and, therefore, it would have been counterproductive to their goal.

Second, another objective of teaching the Holocaust was to engage students' values and to formulate lessons they could apply to their own lives. The disciplinary approach specifically focused on the historical particularities of a single event, making any connections to contemporary events difficult. In retrospect, the disciplinary approach would have been the least controversial, because it inherently recognized the historical uniqueness of the Holocaust. But in the socio-curricular climate of the 1970s, teachers considered historical "relevance" more important than developing historical skills.

Third, an underlying assumption of the disciplinary approach was that historical knowledge is contested and uncertain. In the context of the rising Holocaust denial movement of the 1970s, suggesting that certain facts of the event were contested and uncertain could have potentially been viewed as a form of Holocaust denial. Many educators were teaching the event to combat the Holocaust denial movement. For these reasons, the disciplinary approach was not employed by any of the teachers in these studies until the USHMM teaching guidelines. However, the disciplinary approach was the foundation for many of the critiques hurled at the more progressive-oriented curricula.

In the late 1980s and early 1990s, the growing concern with transmitting the facts of history and engaging students in "historical thinking" led many researchers to stress the particularities of history as a way of complicating students' assumptions. Emphasis shifted from using the Holocaust for comparisons and generalizations to studying it for its unique aspects. The USHMM teaching guidelines reflected this growing concern. But even when educators began stressing the historical and definitional uniqueness of the Holocaust, they did not go so far as to assert its metaphysical uniqueness, as Jewish particularists had done.

In summary, the objective of the traditional approach is to transit a body of content to students. The goal of the disciplinary approach is to have students understand how history is constructed. The objective of the progressive approach is to draw upon all the disciplines to explore problems in the present. These three approaches can help us to understand the scope and manner of Holocaust education, but not necessarily its overall aim, objective, or purpose. In fact, most of criticisms in the study have pointed at the method of instruction, as if the method inherently dictated the outcome. Simone Schweber's counterintuitive study of the effectiveness of the Holocaust simulation reminded us that there are various ways to arrive at the goal of deep understanding.

Before we start sorting through which paths are more desirable than others, we must first arrive at a fuller understanding of exactly what Holocaust education is trying to accomplish at the cognitive and moral levels. Objectives such as "deep understanding," "transformation," "critical thinking," "critical literacy," and "historical thinking" can be as vague and as the superficial as ones they hoped to replace, if they are not directed toward specific and measurable ends. What is needed is a framework that includes both moral growth and disciplinary understanding—that addresses both the familiar and the strange of the Holocaust without necessarily pitting them against each other. To this, we need to revive the cognitive developmentalism of Lawrence Kohlberg.

Kegan's Orders of Consciousness

Kohlberg's theory of cognitive developmentalism served as a major impetus and rationale for the cognitive revolution, which helped to launch Holocaust education in American schools. Somewhere along the way educators largely abandoned his ideas. Part of this can be attributed to the Harvard psychologist Carol Gilligan's attack on the theory. Gilligan argued that his theory was prejudiced against females who tended to put relationships and care above law-and-order, which led girls to score consistently lower on his scale.[18] In addition, having been lumped together with "values clarification" in the backlash against the affective revolution, many viewed Kohlberg's theory as relativistic. In fact, Kolhberg had designed his theory specifically to combat moral relativism. Although the curriculum designers who drew upon Kolhberg's theory in the 1970s continued to use many of the techniques and activities he suggested, they never followed through on these activities by assessing their students using his scale. This was a major oversight.

Reviving Kohlberg's theory in a slightly expanded form can help to bring purpose and coherence to Holocaust (and history) education. It will enable us to resolve the conflicting objectives and methods of the traditional, disciplinary, and progressive approaches to history. To do so we need to look at the work of Robert Kegan, who has expanded the cognitive theories of Piaget and Kohlberg in both breath and length.

Unlike Gardner and Wineburg's disciplinarian cognitive psychology, which suggests that the disciplines create and maintain their own domain-specific symbolic systems and heuristics, Kegan argues that the mind can be viewed in terms of overall orders of consciousness that structure the processing in all domains. The orders are, as Kegan explains, "not merely principles for how one thinks but for how one constructs experience more generally, including one's thinking, feeling, and social relating."[19] The orders relate to the *organization* of thoughts, feelings, and social relations, but not the content of these areas. Each level of mental organization has its own inner logic (or "epistemologic"). Subsequent levels of consciousness do not merely build upon the former, but instead incorporate it. So the new mental scheme is not just the accumulation of specific skills in the behavioral/empiricist sense, but instead represents a transformative, qualitative change. Most significantly, as Kegan argues, "Each successive principle 'goes meta' on the last."[20]

Kegan's theory relates back to Piaget's notion of subject–object relations. Transforming the epistemologies of the mind, he argues, "liberating ourselves from that in which we were embedded, making what was subject into object so we can 'have it' rather than 'be had' by it ... is the most powerful way I know to conceptualize the growth of the mind."[21] In other words, the mind develops an organizational scheme to think *through* and *with*, but upon growth to the next order, it then thinks *about* that very scheme—that is, the mental scheme goes from the subject of thought, feelings, and social relations to the object of thought, feelings, and social relations.

Kegan extended his theory beyond the scope of Piaget's work to include further development that educated adults may experience later in life. Overall, a fully developed adult (of which there are very few) would pass through four orders of consciousness to arrive at the postmodernist mind. Individuals at first-order consciousness (i.e., young children) deal only in immediate perceptions and impulses. The mind is perceived as the only subject in the universe and all sensations (objects) relate directly to it.

Individuals at second-order consciousness can construct what Kegan calls durable categories. They can perceive of themselves as having enduring tastes, likes, and needs, and they can comprehend cause-and-effect relationships between external objects. They use their durable point of view (subject) to process their sensations (objects). Individuals at third-order consciousness view the world through cross-categorical relationships. They understand themselves in relation to how others perceive them, and they can develop mental abstractions of interpersonality such as shared ideals, generalizations, and values. They use interpersonalism (subject) to process their own point of view as well as the perspectives of others (object).

Those at fourth-level consciousness perceive the world through complex systems. They can comprehend relationships between abstractions and can employ an ideology of a discipline and/or institution. They use the discipline (subject) to process the interpersonalism of themselves and others (object). The disciplines of history, sociology, and psychology all use a different, but internally consistent, interpersonal set of tools to process discrepant information. Finally, those who reach fifth-level consciousness can think trans-systematically. They understand the limitations of the discipline or institution and how the rules of such have penetrated themselves and others. They use interdisciplinary understanding (subject) to process disciplines (subject). As you can see, in Kegan's theory the subject of prior level becomes the object of the next.

Overall, Kegan argues that the evolution of the mind works through a process of differentiation and then integration. One "continually sees that a critique of one's identification with the values and loyalties of one's cultural or psychological surround precedes the construction of [the next] order system that can act upon those values, set them aside or modify and reappropriate them to a new place with a more encompassing organization."[22] Kegan suggests that the objective of education, as Dewey suggested over a century ago, is growth. However, we must make a distinction between Dewey and Kegan's theories. Dewey designed his educational philosophy specifically to overcome the subject–object divide, which is at the heart of Kegan's theory. But what Dewey meant by objects were external perceptions, and what he meant by subjects were a priori mental structures. He argued that experience united these two entities. Kegan also unites the two by suggesting that perceptions, thoughts, and feelings are not inherently subjects or objects, but can be either and/or both, depending on one's order of consciousness. In this sense Dewey and Kegan are in accordance. But, Kegan's orders of consciousness are far more specific and useful than Dewey's vaguely defined "growth," because the objectives of the former can actually be identified and measured.

Orders of Consciousness and Holocaust Education

What does all of this have to do with Holocaust education? If we reexamine the three approaches to teaching history we see that traditional orientation is at the level of second-order consciousness, because it focuses on the transmission of durable facts and narrative. The objective of the traditional strand in relation to the Holocaust is either conveying historical lessons, commemoration of the event and its victims, or cultural literacy. In the former case, philosopher George Santayana's oft-quoted dictum that "those who do not know their history are destined to repeat it" was commonly used as a rationale in the early years of Holocaust education. In the 1960s and 1970s, when the event was largely unknown, proponents of Holocaust education simply wanted to move the event onto the radar screen of Americans, especially in light of contemporaneous genocides taking place. If people were aware of the magnitude of the Nazi genocide, early proponents hoped, then they would be more likely to prevent contemporary ones.

Once the Holocaust became better known, educators argued how comprehensively teachers needed to cover the event and what elements they needed to explore. Critics outlined the many important facts and elements of the Nazi campaign against the Jews that were being ignored and overlooked. Rarely did these suggestions and critiques come with clearly articulated rationales—coverage for its own sake with an implicit moral imperative to do so was usually the goal. Accuracy and depth of understanding, it was believed, better commemorated the victims.

Cultural literacy offered another rationale for learning the facts of the Holocaust, although it didn't necessarily clarify what minimal amount of content needs to be covered. Hirsch's *Cultural Literacy* text, which was published in 1987, listed only seven Holocaust-related entries: Auschwitz, Dachau, Adolf Eichmann, Adolf Hitler, Holocaust, Nazism, and Gestapo (Anne Frank, Elie Wiesel, *Kristalnacht* were not listed). Recall that one of the main purposes of having cultural literacy was to increase reading comprehension. With this goal in mind, knowledge could be superficial. For example, to comprehend the text of philosopher Theodore Adorno's well-known dictum that "there can be no poetry after Auschwitz" (his exact words were somewhat different), one would only have to understand that Auschwitz was a representation for the Nazi persecution of the Jews, not necessarily the exact function, location, or history of the notorious death camp. Furthermore, if one were to substitute Hitler, Eichmann, or Dachau for Auschwitz in Adorno's assertion (i.e., "there can be no poetry after Dachau"), then the meaning of the sentence would be the same, other than a slight variation of degree.

No matter what terms one designates as necessary for cultural literacy, the objective, according to Hirsch, was simply to recognize the terms and generally know what they meant. So one would not need to know, for example, that Dachau was a concentration camp and Auschwitz was a death camp, but simply that both were related to the Holocaust. However, knowing the difference between these two camps was indeed a distinction deemed important (presumedly for commemorative reasons) by both Henry Friedlander in 1979 and Karen Riley in 2001.[23] Both authors complained that teachers were not making this distinction. Ultimately, everyone agrees that some concrete factual knowledge of the Holocaust is necessary, but exactly how much information needs to be included to satisfy the demands of commemoration or cultural literacy remains a mystery.

The disciplinary and progressive approaches are both more meaningful than the traditional. Although drawing on different disciplinary traditions (history and social science), both the progressive and disciplinary strands aim to move students to third-order consciousness. Specifically, the disciplinary orientation relates to a relationship between the reader–historian (subject) and the historical document–text (object). The objective here is to appreciate the "strangeness" of the document by considering its subtext. To do so, as Samuel Wineburg had demonstrated, one has to abandon the idea that a document can be read as direct window into the past, but instead it must be read as a point of view.[24]

Overall, the goal of Holocaust education under the disciplinary perspective is to create a deeper level of understanding of the past by appreciating the limitations of our present understanding. At a higher level of consciousness (i.e., advanced undergraduate and graduate school), the disciplinary approach constructs complex procedures for constructing and assessing the value of histories to arrive at fourth-level consciousness. Beyond that (i.e., graduate school, specialized historiographical study) the disciplinary approach aims to have students appreciate the limitations of relying on the forms and methods discipline and its theories to arrive at fifth-level consciousness.

No issue underscores the significance of the disciplinary approach more than the idea of anti-Semitism. As Simone Schweber asserted, anti-Semitism was "the definitive root cause for this genocide to have occurred and a key exemplary framework for understanding the Holocaust."[25] The racial form of anti-Semitism that existed in Europe during the first half of the twentieth century does not exist in America today (at least not to the same degree). Trying to get students to understand this condition is a central concern for those in the disciplinary camp, for it represents the primary explanation for why the event occurred. This is not just a matter of transmitting the facts of anti-Semitism, but involves developing empathy for those in the past by appreciating how the historical participants viewed the world differently than we do now. Such a process presupposes third-order consciousness.[26]

On the other hand, the progressive approach to history education links the past to analogous examples in the present. Such an effort is intended to make the content immediately relevant and result in moral and civic transformation. Although the disciplinary orientation has moral elements, the progressive orientation is more explicitly moral because it relates to relationship between citizen–actors and their ethical systems instead of historical texts. To shift students from second- to third-order consciousness, students must appreciate that other citizens may have different value systems, even if they share a core of common beliefs. They must appreciate cross-categorically that the Germans in 1930s had specific circumstances and beliefs; they were not just evil. At the fourth-level of consciousness, students (i.e., advanced undergraduate and graduate school) can appreciate and critique the socially constructed nature of moral systems including their own.

In the last chapter, we explored examples of college-age Christian students who were shocked that their existing moral system did not explain the German anti-Semitism they encountered in the course. Having experienced cognitive dissonance they were forced to construct a new cognitive paradigm to explain and incorporate the dissonant information. Some may have resisted and withdrawn into the comfort of their existing moral system by dismissing their instructor as being "anti-Christian" or "secular." But the successful students moved from a second-order to third-order consciousness by developing a cross-categorical construction between his/her own beliefs and those of the anti-Semitic Christians of the past. This new consciousness would hopefully be transferable to other moral and historical dilemmas.[27]

The tension that has appeared throughout this entire narrative has been the alleged incompatibility of relating to the past and relating to the present, of understanding the relevance of the Holocaust through analogous examples in the present, and appreciating its historical particularity by recognizing the strangeness of the past. This divide works on the assumption that one must either appreciate the past on its own terms, or make it relevant to today. However, the cross-categorical understanding of third-order consciousness requires a two-way street between the past and the present. It requires an appreciation of one's own moral system and historical milieu as a precondition to understanding those of another. A student at second-order consciousness would work through the durable categories of the present, and be unable to appreciate the durable separate categories of the past. Such a student would likely say that the Nazis were simply stupid and evil. To understand that people in the past felt differently, the student would simultaneously have to "go meta" on him/herself in order to appreciate to some degree the strangeness of his/her own moral system.

What I am suggesting is that the two sides, progressive and disciplinary, are not only compatible but are in fact dependent upon each other. One cannot have one without the other, because one must become conscious of his/her own moral and disciplinary perspective, before appreciating how another's may be similar or different. Moral growth and growth of historical understanding are related and coterminous. They both involve reconciling the categorical understanding of self with those systems that seem to undermine it. They both require the development of cross-categorical understanding. In the mental transition from second- to third-order consciousness, moral and disciplinary understandings develop simultaneously and symbiotically.

Reviving Kohlberg: Development as Aim in Holocaust Education

Although many of the earliest Holocaust curricula used Kohlberg's theory as rationale for Holocaust education (and some instructed teachers to introduce the theory directly to students) they did not really apply his theory to the construction of their objectives or their assessment. What I am suggesting is that public school teachers and researchers use the orders of consciousness as the main objective of Holocaust and secondary education. While commemoration may be a valid rationale for Jewish teachers in Jewish settings, and cultural literacy may be achievable for public schools, I do not believe these are appropriate overall goals for Holocaust education. This means if a student does not know the difference between Auschwitz and Dachau—so be it. These facts are only useful so far as they will help students achieve cognitive growth.

Studying the Holocaust for cultural literacy is equally problematic, because it is unclear what information to include. As we have seen repeatedly throughout this narrative, Jewish Americans are likely to have different ideas than non-Jewish Americans about the minimal amount of the content that should count as cultural literacy. More importantly, both these rationales engage students at the second-order of consciousness, but make no attempt to move them further. The facts (or specific interpretations of them, to speak at the level of third-order consciousness) are significant and necessary, but should not be viewed as an end in Holocaust education.[28]

However, I am not suggesting that students have to first "know the facts" in order to move to the next level, because teachers can assume that by middle and high school, students are already at the second-order consciousness. In other words a teacher does not need to recapitulate the orders of consciousness for each new subject by first teaching them the facts. I am also not suggesting that because the objective of Holocaust education is growth, any topic or genocide can be substituted for the event as long as it contributes to student growth.

The Holocaust is an extremely important event in its own right, but this importance has emerged over the years in relation to how Americans have responded to it in a variety of cultural ways. To use a Deweyian, pragmatic view of the event, the significance of the Holocaust made itself apparent over time; its significance was not, as many particularists have asserted, inherent in the event itself. Ultimately, the Holocaust should be taught to American students because it has become one of the most important historical topics of the twentieth century. It has impacted the way foreign policy and international relations are conducted, the way modern warfare is conceived, and the way in which ideas such as "modernity" and "rationality" are viewed. It clearly has scholarly, cultural, and moral significance for all Americans.

So overall, the traditional approach to the Holocaust offers a weak rationale for learning about the event, and will not lead inherently to moral and cognitive growth. On the other hand, both the disciplinary and progressive approaches to history are inherently moral, and both require shifting from the second- to third-order consciousness. This is an achievable goal for middle- and high-school students. The research of Bruce VanSledright with fifth-graders demonstrates that even students at that age can engage in this type of thinking.[29] However, Samuel Wineburg's research has shown that successful Advanced Placement high-school students will not automatically reach third-order consciousness, if not explicitly taught to do so.[30]

However, I suspect that orders of consciousness in the areas of moral and disciplinary thinking are related to each other, but are to a certain degree distinct. That is, I do not believe that reaching a third-order consciousness in one domain leads to automatic growth in the other, but it probably makes it easier.[31] The following examples support my assertion.

In Simone Schweber's study of the Holocaust simulation, she interviewed a student participant named Calypso, who related how she was so dedicated to Ms. Bess's class that she came to school just to participate. Calypso's attendance was not just linked to her grade in the course, but, more significantly, it was linked to the survival of her imaginary Gestpo simulation character and her cherished ones. "I woke up at like twelve," Calypso explained, "and I was like 'Damn I need to get to Ms, Bess' class!' And I just threw on a tee-shirt and jeans just for that class. And then I didn't even go to track practice; I just went home! And then, . . . I was like '*Oh God, I'm like one of those kids,' and I was like, Na ah*'" (italics added).[32] Calypso was surprised by her own dedication to her character and cherished ones whose lives she risked by not coming to class. She had established a cross-categorical relationship with these individuals from the past, and she defined her own identity in relation to them. This is an example of Calypso "going meta" on her own identity. She was shocked at the transformation from a somewhat apathetic student to "one of those kids"—a transformation she was not completely comfortable with. She was transforming her identity into something that *she had* rather than something that *had her*. She was in the process of moving from the second- to third-order consciousness.

Another example from Schweber's ethnography exemplifies such growth. During the simulation, a discussion emerged about whether it was morally justified to sacrifice the lives of others to save your own. One student Adrienne refused to be selected as a *Kapo*, a role that would have possibly involved selecting others for extermination. However, another student, Thomas, took the job with little hesitancy. When Ms. Bess asked Adrienne why she rejected the job, she replied that it was because she did not want to have to kill people. In response, Thomas explained his position; if his own life was threatened and he was going to be killed if he did not obey, he would do anything to survive. After different students chimed in as to whether they would side with Adrianne or Thomas, one student astutely observed that the students' positions tended to fall along gender lines.[33] Once again, through this discussion we can see the students going meta on their own moral systems, which are shifting from a cognitive organization they think *through* to one they think *about*. Their moral stances are shifting from the *subject* of their perceptions, feelings, and social relating to the *object* of their perceptions, feelings, and social relating. This is the kind of transformation to which Holocaust education should be aimed—one in which understanding of historical context mingles with and is challenged by one's identity in the present.

The orders-of-consciousness scheme was not meant to be critique of teachers of the Holocaust, but in fact to defend them. There are numerous documented examples of teachers achieving remarkable things in their classrooms, but the teachers do not necessarily have the vocabulary to articulate exactly what they are doing. For example, Melinda Fine's 1993 ethnography of *Facing History* curriculum focused on the confrontational student Abby. As Abby explained to Fine, "I think that the right thing should always be promoted and the people who think otherwise should not be allowed to speak." Abby was clearly working at second-order of consciousness, and her teacher was trying to move her to the third-order when she responded, "Abby, there's a bit of tunnel vision here that we have to speak about. You can't just say that the only ones who can speak are those who agree with your position!" Appropriately, Fine used this quotation in the title of her article to capture the fundamental goal of *Facing History*. The teacher may not have been immediately successful at shifting Abby to the next order of consciousness, but she had clear conception of what she hoped to accomplish.

Nazism and Cognitive Developmentalism

By now the reader might be wondering if the orders of consciousness is just another name for moral relativism. Could the Nazi point of view actually be justified by using such a scale? Do we really want to teach our children to be moral relativists by tolerating intolerance? Of course not. The Nazi ideology is a closed system constructed at the level of second-order consciousness. The Nazis constructed a racial ideology that served as a durable category, a single egocentric epistemology that did not include any cross-categorical understanding. That is, the Nazis constructed their worldview upon a categoryical set of "truths," which did not in any way include the views of those outside it. All other viewpoints were literally outlawed and persecuted. The most obvious and catastrophic example of this was how the Nazis viewed the Jews as subhuman, a viewpoint that clearly did not consider how the Jews viewed themselves or how the Jews viewed the Nazis.

This is not to dismiss the Nazis as products of poor cognitive development. I agree with philosopher Berel Lang, who suggested that the Nazi leadership was fully aware that they were committing evil when they implemented the final solution. "One preeminent feature of the final solution," Lang explained, "is that it was intended to be concealed from beginning to end... Great efforts were made to obliterate traces of the existence of the death camps: to erase traces of the gas chambers and crematoria that they employed; of the bodies or graves of their victims; and of the identities and the numbers of the victims."[34] As adults, the Nazis were capable of achieving a higher-order consciousness, but deliberately made efforts to avoid doing so. In fact, as Lang argued, the process of dehumanizing the Jews before exterminating them was not only a sinister sadistic act, but a necessary precondition to genocide. That is, the Nazis had to take carefully planned steps to avoid going meta on themselves—to keep themselves firmly embedded at the level of second-order consciousness. "The process of systematic dehumanization," Lang argues, "requires a conscious affirmation of the wrong involved in it—that is, that someone who is human should be made to seem, to become, and in any event be treated as less than that."[35] In other words, the Nazis needed to prove empirically the subhuman nature of the Jews through humiliation as a false justification for murdering them—a process that inherently recognized their victims' humanity.

In a 1942 address, John Dewey also addressed the shortcomings of the Nazi's totalitarian ideology. First, Dewey argued, Nazism suggested, "the basic truths by which society ought to be governed are primarily in the possession of a relatively small group, which in virtue of their possession of it, are morally entitled to be what the Nazis called a 'leaders' of the mass." Second, Nazism asserted that "the great mass of people cannot be trusted to arrive at possession of these truths... [without those] whose monopolistic possession of ultimate social and moral truths gives them the responsibility as well as the right to leadership." What troubled Dewey about the antidemocratic means of the Nazis was the fact that "so many firmly believe that it is used for high social ends, and its victory will contribute positively to the order."

However, in 1942, Dewey seemed optimistic that the Nazi project would not succeed because "When rule is found to rest merely upon superior force its days are numbered. In order to be able to control the lives of men, a ruling power has to clothe its power with authority, with what is at least a semblance of right."[36] Dewey suggested that principles of truth should rest upon the combined experience of all, not the absolutes of a particular imposing group. Such morality, whether it be the Christian fundamentalism with which Dewey grew up or the Nazi domination he observed from afar, is empty (or what he called a "colorless innocence") unless it is confirmed in the experiences of all, not some.

Perhaps far more simply, the Nazis violated the inalienable individual rights of human dignity. The authors of the affective revolution understood that a belief in human rights and democracy ultimately came down to faith. As Donald Oliver explained, humanity must be committed to promoting the dignity and worth of each individual. "There is not final proof of such value; when one pushes to the heart of human values, he must invariably end up accepting some tenet on faith."[37] Likewise Lawrence Kolhberg explained that his developmental scheme is ultimately empty unless it moves toward "a philosophic notion of adequate principles of justice."[38] I believe such a foundation of faith in the dignity of every human being, whether it be spiritually based or secular, is a necessity in dealing with the moral complexity of the Holocaust.

Uniqueness and Teaching the Holocaust

Of course, no one can fully comprehend what it was like to be a Holocaust victim, but to suggest that one should not even attempt to understand how and why the event occurred runs counter to the entire post-Enlightenment rationalist tradition. Teachers cannot, as Elie Wiesel suggested, treat the Holocaust as a metaphysically unique event. Samuel Totten, recognized this point, writing "While I greatly appreciate Elie's work, I think that his position is one that leaves educators, philosophers, theologians and others no place to go or do other than weep endlessly."[39] Likewise, Simone Schweber suggests that particularist views of history "are not dedicated to the learning of history as a discipline but rather to the inculcation of Judaism as a 'total world.'"[40] Although many teachers in these case studies used the writings of Wiesel as a rationale for teaching the Holocaust, or used his fiction in their instruction, none of them have adopted his metaphysical views on the Holocaust.

Regarding Holocaust uniqueness, these case studies have demonstrated that it is virtually impossible to retain all the elements of uniqueness, and still design a curriculum through one of the social studies approaches. To make the topic resonate with students of a diverse background and make it applicable to current events, the historical uniqueness of the event must be compromised. Even when curriculum designers, like those of *Facing History*, clearly asserted the definitional uniqueness of the Jewish experience, their mere inclusion of other events and prejudices inspired criticisms and objections.

The destruction of the European Jews by the Germans is a historical event that took place in time and space like any other. The Holocaust is indeed unprecedented as a phenomenon; it was the only time a Western nation has used its full bureaucratic and military force to exterminate an entire group of innocents simply for belonging to what they believed to be a particular race. This unprecedented campaign is the worst example of genocide in history. However, this is a warranted assertion, not an epistemological position. To suggest that the Holocaust requires a unique set of understandings, or that this particular event should be approached with a unique reverence, is inherently disrespectful to other calamities, atrocities, and mass killings of the past and present. I agree with the USHMM teaching guidelines that teachers should avoid comparisons of pain. This position recognizes the uniqueness of each atrocity and each victim.

However, to follow this uniqueness-of-each-event claim through to it logical conclusion, whatever unique reverence had been suggested for the Holocaust should also apply to all other historical events involving death, lest a hierarchy of suffering unavoidably be established. This means that crossword puzzles about Hiroshima should be off limits, and multiple-choice questions about Alabama church bombings should be abolished. In addition, the kind of Civil War battle reenactments in which park rangers routinely lead visiting students should be outlawed. Films about slavery should not be made, and books about fictional terrorist plots should not be published. To some, these suggestions are absurd. Yet if we are to treat every historical man, woman, and child who ever suffered with the respect and dignity afforded the Holocaust victim (a respect, I admit, they clearly deserve), history would be nearly impossible to teach. The voice of each victim would need to be heard, and the full story of every atrocity would need to be related in full. To be consistent, teachers must either treat all victims as definitionally, historically, and metaphysically unique, or treat none of them that way. Pedagogically speaking, the latter seems far more realistic.

Improving Holocaust Education

Rather than endorse a particular approach to teaching the Holocaust, I have chosen to outline the strengths and weaknesses of each; no curriculum is perfect. Teachers must have the freedom to choose a curriculum that best meets the needs of their particular student body and their particular teaching style. They must also have the freedom to tinker with the materials in a way that remedies its perceived weaknesses. The curriculum pioneers covered in this study understood this point. They worked with teachers through workshops and instructional materials and allowed curricular flexibility. This is why Holocaust education has been one of the most successful grassroots education movements in American history. That being said, teachers should be cognizant of the trade-offs they are making in their curricular choices.

As I have mentioned, effective teachers often employ elements of the traditional, disciplinary, and progressive curricular approaches to history in their instruction; the first teachers of the Holocaust did the same. If my narrative has drawn too strong a distinction between these individual approaches (and curricula), this was done in order to emphasize the potential consequences of each. But these strong distinctions, whether they existed or not, were indeed made explicit by the various critics of Holocaust education themselves, who often characterized these curricula as being more extreme than they actually were. They polarized these approaches to make their own position seem clearer. In the end, the debate over the Holocaust curriculum was and is a natural consequence of trying to teach all citizens in a pluralistic, democratic society. Through these debates, the strengths and weaknesses of each approach are brought out.

The cultural divide between historians and teachers is at the heart of the controversy over the Holocaust curriculum, and both are to blame. Since the beginning of the century, historians have cut themselves off from the secondary school curriculum. In *History on Trial*, Nash, Crabtree, and Dunn argue that after the founding of the social studies, historians took a "long walk" away from direct interest in the schools, only to return in the 1990s with the collaborative (but politically divisive) work on the national history standards.[41] Nash et al. overstate their case, as Holocaust historians in this study have demonstrated. Leading Holocaust historians such as Henry Friedlander, Raul Hilberg, Lucy Dawidowicz, Yehuda Bauer, and Deborah Lipstadt have voiced their opinions on how and why the Holocaust should be taught.

Unfortunately, often the historians' critiques demonstrated their ignorance of what goes on in the classroom. Deborah Lipstadt in her critique of *Facing History* suggested that students would be able to relate the lessons of the Holocaust to contemporary issues without making the connection explicit, because *her* students could do so. Comparing her college-age students at Emory, one of the nation's top universities, to the ethnically diverse, urban eighth-grade students of Brookline, Massachusetts, is insensitive and absurd. It demonstrates how scholars often have little to no knowledge of how difficult it is to interest students, who have not grown up in middle-class, suburban white neighborhoods and have no intention of going to college, in history. But at least these historians have taken interest in the topic. Have leading historians of the Civil War or Civil Rights movement ever written critiques of how these events are taught in American classrooms? Do they even care? Shouldn't the meaning of the Civil War and Civil Rights movement be as important to the formation of American citizens as the Holocaust? Reflection on the Holocaust curriculum raises these important questions.

Of course, teachers are not completely innocent. When leading historians of a particular topic are writing that certain curricula are historically inaccurate, teachers should take heed. When students finish a Holocaust unit thinking that a Holocaust could break out in America at any time, or, as one student reported, "there's an Adolph in me and an Adolph in you," they are learning that historical context is insignificant. They are learning that prejudice is ahistorical. Admittedly, Kohlberg's scale of moral reasoning encouraged such ahistoricism. Nevertheless, this kind of conclusion is irresponsible and wrong. Not only does it fail to recognize the historical particularities of the Jewish victims of the Holocaust, but it also inherently disrespects the historical particularities of other genocides, atrocities, and prejudices—most notably American slavery and racism, subjects that are just as relevant to American students as the Holocaust. Students should learn about the historical particularities of all these events in their full historical context. Here, I have suggested how moral growth and historical empathy can and should develop simultaneously.

Holocaust Education and the Aims of Secondary Schooling

Ultimately, the Holocaust is not a discipline, it a subject, a collection of evidence. But there is no better subject in the world to develop moral and cognitive growth. Not only should Kohlberg–Kegan shifts in orders of consciousness be the objective of Holocaust education, history education, and secondary education in general, but it should also be the means of assessing to what degree Holocaust education was a success in the long term. In this manner, understanding the Holocaust becomes a means, not an end. The rich body of empirical research on Holocaust education has demonstrated how complex the endeavor can be.

Educators must accept that spending a few days or even weeks teaching the Holocaust is not going to transform anyone. Spending an entire semester on the Holocaust can be more meaningful, but even then, transforming the embedded cognitive structures of the mind is a long, slow process, and it results from numerous encounters both inside and outside of the classroom. For this reason, Holocaust education needs to be viewed as a component of the overall goals of secondary education, not merely as an end in itself.

While Holocaust education experienced unprecedented growth in American middle and high schools during the 1980s and 1990s, during this time the institution of comprehensive secondary school itself came under attack from various educators, politicians, and policy makers.[42] Barry Franklin and Gary McCulloch suggest that there are three core questions underlying these controversies. The first addresses who the high school should serve; should it provide an education for all adolescents or should multiple institutions serve some youth and not others? The second addresses what the high school should offer; should it provide a common educational experience for all adolescents or simply serve the unique needs of particular youth? The third, and most significant to Holocaust education, addresses the overall purpose of the institution; should the high school be aimed toward developing democratic citizens, training the nation's workforce, or securing the social mobility of adolescents? For a century, proponents and defenders of the comprehensive high school have ambitiously asserted that it should accomplish all three of these objectives. Opponents have asserted that such comprehensiveness has led to a "shopping mall" curriculum.[43]

As we have seen, the origins of Holocaust education was a direct product of the shopping mall structure of the secondary school of the 1970s and 1980s. Accordingly, many of the first teachers of the Holocaust addressed the event not in required courses, but in voluntary, elective ones. Many still do. However, once the Holocaust education became popular, and evidence emerged of its effect on adolescents, certain politicians and policy makers sought to make it mandatory. These initiatives were very much in line with the movement toward a standardized curriculum because educational leaders believed that it would be undemocratic to deny knowledge of the event to certain students. As a result the topic of the Holocaust was thrust upon certain teachers who were not necessarily qualified to teach it. At present policy makers in individual states have taken a range of positions on whether all students should learn about the Holocaust, or whether learning about the event should be left to choice or happenstance (i.e., the fortune of having access to a teacher with a Holocaust profile). In this sense, the history of Holocaust education at the state level is still being written.

Regarding the aims of secondary education, teaching the Holocaust seems to be focused primarily upon the democratic objectives of developing civic and moral virtue in all students. As we have seen, there is evidence that students of different economic, social, and ethnic backgrounds have benefited from learning about the Holocaust, so the movement is not aimed at socializing students into middle-class capitalist culture, or as a means of social control. Nor does Holocaust education seem to be connected narrowly with preparation for college or for acquiring credentials for particular professions. Instead Holocaust education in America arose from successful teacher-led grassroots effort to reestablish the democratic function of secondary schools—to explore a controversial topic with all students, regardless of class, ethnic, or academic background, that expands their intellect, inspires their character, and challenges their values. Such should be the overall goal of secondary schooling.

Epilogue

The Future of Holocaust Education

Content about the Holocaust is now firmly embedded in the American curriculum. Nearly every major city has some kind of Holocaust museum that creates and distributes educational materials. For more than a decade, the USHMM has trained hundreds of teachers through its summer internships and teacher workshops, and over the years the Facing History and Ourselves Foundation has trained thousands of teachers and distributed educational materials to many more. In addition teacher guides and educational materials are often created to accompany Holocaust commemorative and popular events—most recently the 2005 PBS documentary on Auschwitz.[1] The research presented in chapter seven demonstrates that the Holocaust is being taught in a number of creative contexts and through a diverse range of approaches throughout the country. In addition textbooks cover more material about the Holocaust than ever before, and those states with history standards often list the Holocaust specifically. The Holocaust as a topic in American schools is healthy.

However, the kind of Holocaust education described in this study is endangered. Unfortunately, in-depth investigations of any historical topic are becoming more difficult in the age of standardized testing. Recent research shows that in elementary classrooms social studies is being pushed right out of the curriculum to make room for remedial and review instruction in reading and math, subjects that are tested in accordance with the No Child Left Behind Act.[2] At the secondary level, many states that do have history standards listing the Holocaust often emphasize superficial coverage over in-depth investigation.[3] In one eleventh-grade classroom in Virginia, I witnessed the Holocaust being taught in only three minutes, as a series of bullet points on a PowerPoint presentation on the effects of World War II. More and more, the content of standardized curriculum and tests dictate what is being taught in American classrooms. In some districts, teachers are told explicitly what content to cover every single day of the year.

In contrast, many of the Holocaust courses covered in my account were taught as electives. While electives continue to be taught, they are becoming far less common in the age of the one-size-fits-all curriculum. However, in the 1970s, social studies elective courses were common. Teachers were encouraged to teach topics like the Holocaust that inspired them and sparked their students' interests. If this seems like a misguided notion, consider the impact of one of the Holocaust courses covered in this study. In the summer of 2006, I received an email by a man who identified himself as a "federal bureaucrat." Having read an article I wrote on the origins of Holocaust education, he wrote the following: "I asked my wife, a seventies graduate of Teaneck High [in New Jersey], if she remembered Edwin Reynolds. She remembered him very clearly and proceeded to give me a very lively description of his teaching about the Holocaust. As far as I can remember for 26 years of marriage, this is the only High School class she has any recollection of!"

Mr. Reynolds clearly provided this student with an engaging, enthusiastic course that stayed with her for decades. His instruction derived from an in-depth investigation into a complex topic, not a superficial race through a catalogue of names and dates. Mr. Reynolds along with Albert Post, Richard Flaim, William Parsons, Margot Stern Strom, Roselle Chartock, Rabbi Raymond Zwerin, Carol Danks, and Leatrice Rabinsky were highly qualified teachers, who demonstrated initiative, creativity, and intellectual leadership. These are the kind of teachers we want teaching our children. These are the kind of teachers we want to attract and retain.

As this study has shown, meaningful investigations of the Holocaust, or any topic for that matter, are dependent upon two things: the knowledge and interest of the instructors who teach it, and a curricular context that supports their creativity and experimentation. To those who wish to increase instruction on the Holocaust, lobbying for state mandates may achieve a rewarding political victory, but this will not necessarily result in the kind of meaningful experience described here by Mr. Reynold's former student. In fact, mandating the teaching of the Holocaust undermines the entire manner in which the movement emerged. Initially Holocaust education was not something imposed from above, but rather something that was developed from below. The movement began in the hearts and minds of a few ambitious teachers who wanted to use history to transform the lives of their students. It grew out of an environment that allowed highly qualified teachers the freedom to experiment, innovate, research, reflect, and share. The future of meaningful Holocaust education is dependent upon such an environment.

This brings us to the paradox of modern educational reform. While state departments of education continue to increase the requirements of secondary teachers entering the profession, they also increase their grip on what is being taught in classrooms through standardized testing and state mandates. In other words, many legislators are trying to recruit and retain the most highly qualified candidates to the profession—presumably those individuals with qualities of initiative, leadership, and critical intelligence—yet they are creating an environment that undermines these very qualities. No wonder half of all teachers will leave the profession within their first five years.

Policy makers must choose one or the other. They must decide whether American secondary schools should be like colleges and universities—where creativity, intellectualism, and innovation are rewarded—or like the military, where uniformity and following orders are valued above all. American education has straddled these two worlds for a century, never fully committing to one or other. When the ambitious teachers of the Holocaust covered in this study retire, will there be any to take their place? The answer to this question is dependent upon the kind of educational environment we create for them in our secondary schools.

Notes

Introduction: The Story of Vineland, New Jersey

1. Proctor, *Racial Hygiene*, 179.
2. Ibid., 97.
3. Ibid., 173.
4. Goddard, *The Kallikak Family*, 117.
5. Ibid., 101–2.
6. See Gould, *The Mismeasure of Man*.
7. Proctor, *Racial Hygiene*, 99–100. Goddard later recanted much of what he proposed, eventually concluding "Feeble-mindedness (the moron) is not incurable," and "The feeble-minded do not generally need to be segregated in institutions." Quoted in Gould, *The Mismeasure of Man*, 174.
8. Goddard, *The Kallikak Family*, 103–4.
9. This quotation is by Richard F. Flaim in Rosenblum, "State Backs," 4.
10. Ibid. Patricia Alex, "Teaching the Holocaust," *The Record* (Bergen, NJ), August 30, 1994, B1.
11. Brabham, "Holocaust Education," 139–42.
12. The International Directory of Organizations in Holocaust Education, Remembrance, and Research can be found on the United States Holocaust Memorial Museum website, http//: ntdata.ushmm.org/ad/. Many of these organizations existed prior to World War II and merely added the Holocaust to their interests; hence, this number does not reflect centers opened exclusively for Holocaust education.
13. See Novick, *The Holocaust* and Mintz, *Popular Culture*. Novick approaches his research on the Holocaust in America with "curiosity and skepticism," and it shows in his methodological approach and conclusions. Using published and archival sources, Novick focuses on the discourse among American Jewish elites and communal leaders. He argues that they have chosen to use the Holocaust to serve a variety of purposes within the Jewish community: to expand solidarity behind the cause of Israel, to combat a perceived resurgence in anti-Semitism, and to garner financial support from marginal and/or assimilated American Jews. Once they established internal political uses for the Holocaust, Novick argues, Jewish leaders used their prominent position in

American academic, political, and popular culture to introduce the topic to the American consciousness. He writes: "We [the Jews] are not just 'the people of the book,' but the people of Hollywood film and the television miniseries, of the magazine article and the newspaper column, of the comic book and the academic symposium. When a high level of concern with the Holocaust became widespread in American Jewry, it was, given the important role that Jews in American media and opinion-making elites, not only natural, but virtually inevitable that it would spread throughout the culture at large" (p. 12).

Novick asserts that the media and financial resources available to the American Jewish community allowed them to exert a disproportionate amount of influence on the way the Holocaust was received by gentile Americans. As far as why non-Jewish Americans have been so receptive to the Holocaust, Novick expresses his uncertainty, "we don't know for sure what's moved them in this direction" (p. 233). Novick does refer to the proliferation of the Holocaust in American education. He uses this section to express his "doubts about the usefulness of the Holocaust as a bearer of lessons," and cynically dismisses the whole venture, writing "if there are, in fact, lessons to be drawn from history, the Holocaust would seem singularly lacking in them, not because of its putative uniqueness, but because of its extremity."

Alan Mintz is not a historian, but a professor of Hebrew literature. His book consists of four loosely linked chapters on the Holocaust in popular culture. In the first, Mintz provides a concise account (thirty-two pages) of the rise of Holocaust consciousness, describing the "pivotal role of popular culture in spreading awareness of the Holocaust from the Jewish community to the larger American nation." Mintz criticizes Novick for a "lack of empathy for its subject" and for giving "short shrift to the role of popular culture in making the Holocaust an American concern rather than just a Jewish one" (p. 187). So his brief narrative aptly places emphasis on "the power of cultural texts and their diffusion in the form of books, stage plays, movies, and television" (p. 16).

Regarding the teaching of the Holocaust, Mintz incorrectly identifies *Schindler's List* as the "most significant instrument in the diffusion of Holocaust education," ignoring the fact that most of the curricular materials on the Holocaust were in fact designed before the screening of Spielberg's movie. For Mintz the teaching of the Holocaust is just another element of the "Americanization" process. He concludes "the lesson about man's inhumanity to man that should instruct us about hatred and intolerance in all walks of life...in the end, is the American teaching of the Holocaust" (pp. 34–35).

As much as Mintz presents his study as a refutation of Novick's, both studies offer similar approaches. Novick's account mentions all the major cultural events contributing to the rise of Holocaust consciousness in America (the publication of Anne Frank's *The Diary of a Young Girl,* the trial of former-Nazi Adolf Eichmann, the Six Day War, NBC's *Holocaust* miniseries, the opening of the United States Holocaust Memorial Museum, and *Schindler's List*), but Novick interprets these events as driven and shaped by Jewish elites.

Mintz, on the other hand, sees these events as driving forces in and of themselves. Mintz attempts to give them agency by casting them as "cultural texts." But, obviously, these cultural texts were created, endorsed, and distributed by people, and, for the most part, these people were Jewish. Ultimately in regard to Holocaust education both scholars are missing the point; non-Jewish Americans participated not only in the reception of the Holocaust but also in the shaping of Holocaust memory, and they had their own reasons and motivations for doing so. Any diffusion model for the rise of Holocaust consciousness must include the voices of the non-Jewish majority in American society.

14. Brief studies of the history of Holocaust education include Littell, "Breaking the Silence," 195–212; Totten "Holocaust Education," 305–12; Ben-Bassat, "Holocaust Awareness," 403–23. There are brief narratives (one–two pages) of the rise of Holocaust education in higher education in Haynes, *Holocaust Education and the Church* and in secondary education in Schweber, *Teaching History.* On the Holocaust in Jewish education, see Sheramy, *Defining Lessons.*
15. Kliebard, "Constructing a History," 158.
16. Mintz, *Popular Culture,* 33.
17. Lipstadt, "Not Facing History," 26–29.

CHAPTER 1 TELLING THE WAR

1. This summary of the events of the Holocaust is based on "History of the Holocaust: An Overview," in *Teaching about the Holocaust: A Resource Guide for Educators,* (Washington, DC: USHMM, 1995), 27–34.
2. Cole "Intercultural Education," 185, 136.
3. Benedict, "American Pelting Pot," 18.
4. Bohan, "A Rebellious Jersey Girl," 99–115.
5. Baron, "The Holocaust and American Public Memory," 64–65.
6. Quoted in ibid., 66.
7. For a discussion on the statistics of death for World War II, see chapter one in Dawidowicz, *The Holocaust and the Historians.*
8. Cole, *Selling the Holocaust,* 21–36.
9. Lipstadt, "America and the Memory of the Holocaust, 1950–1965," 200.
10. The two most cited early studies are Gerald Reitlinger's *The Final Solution* (1953) and Leon Polikov's *Harvest of Hate* (1954).
11. Quoted in Korman, "The Holocaust in American Historical Writing," 255.
12. Ibid., 260.
13. Jick, "The Holocaust: Its Use and Abuse," 309.
14. Ibid.
15. Holocaust is a Latin-derived word meaning "total destruction by fire," or more specifically a "burnt offering." It is a word with religious/sacrificial overtones, specifically in connection with the biblical story of Abraham and Isaac. In contrast *shoah* translates into English as "destruction" with no religious connection.

16. Garber and Zuckerman, "Why Do We Call 'The Holocaust' The Holocaust?," 197–211.
17. Young, "America's Holocaust," 70.
18. *New York Times,* May 4, 1970, A16.
19. Paul Montgomery, "4,000 at Temple Emanu-El," *New York Times,* April 10, 1972, A17.
20. Young, "America's Holocaust," 70.
21. Marcus, *The Treatment of Minorities,* 24.
22. Ibid., 36.
23. Kane, *Minorities in Textbooks.*
24. Ibid., 76.
25. Friedlander, *On the Holocaust,* 11.
26. Irving Spiegel, "A New Historical Guide on Jews," *New York Times,* October 30, 1971, A5.
27. *New York Times,* May 18, 1970, A2.
28. *New York Times,* May 18, 1970, A2; Spiegel, "A New Historical Guide on Jews," A5; Friedlander, *On the Holocaust.*
29. Sheramy, *Defining Lessons,* 6–11.
30. Ibid., 9–10.
31. Zwerin, interview with author.
32. Quoted in Sheramy, *Defining Lessons,* 32.
33. Ibid., 29.
34. Ibid., 34–35.
35. Quoted in ibid., 43, 44.
36. Zwerin, interview with author.
37. Quoted in Sheramy, *Defining Lessons,* 55.
38. Ben-Horin, "Teaching about the Holocaust," 5.
39. Cited in Roskies, *Teaching the Holocaust to Children,* 5–6.
40. Littell, "Breaking the Silence," 198.
41. Krefetz, "Nazism," 7.
42. Cited in Roskies, *Teaching the Holocaust to Children,* 23.
43. Ben-Horin, "Teaching about the Holocaust," 5.
44. Pilch, "The Shoah" 163, 164.
45. Feinstein, "The Shoah," 165, 166.
46. Ury, "The Shoah," 168, 170, 171.
47. Toubin, "How to Teach the Shoah," 22, 24.
48. Stern, *Learning More about Pathways*; Spotts, *Guide to Teachers.*
49. Charny, "Teaching the Violence of the Holocaust," 15, 16, 22, 23.
50. Blumberg, "Some Problems in Teaching the Holocaust," 13, 14, 16.
51. Quoted in Bennett, "Towards a Holocaust Curriculum," 22.
52. Ibid.
53. Broznick, "A Theological View," 19, 20.
54. Spiegelman, "On the Holocaust," 36–37.
55. Roskies, *Teaching the Holocaust to Children,* 6.
56. Ibid., 45.
57. Stadtler, *The Holocaust: A History of Courage,* xiii.

58. See Mintz, *Popular Culture* and Shadler, "Aliens in the Wasteland," 33–44.
59. Cohen, "Just the Facts," 67, 69.
60. Wiesel, *All Rivers Run to the Sea*, 267–70, 319–21; Franciosi, "Introduction."
61. Elie Wiesel, "Telling the War," *New York Times*, November 5, 1972, Section 7, 3.
62. Ibid.
63. Littell, "Breaking the Silence," 197–98.
64. Fred M. Hechinger, "Holocaust: A Course Created by Students," *New York Times*, February 27, 1972, D9.
65. Ibid.
66. John T. McQuiston, "Could There Be Another Holocaust?" *New York Times*, March 6, 1977, W11. Terrence Des Pres, "Lessons of the Holocaust," *New York Times*, April 27, 1976, A35.
67. Ellen K. Coughlin, "On University Campuses," *Chronicle of Higher Education*, May 1, 1978, 2. Peter Novick in *The Holocaust in American Life* referred to this estimate as "possibly inflated" (p. 188) and I agree. Littell estimated that by 1978 over seven hundred courses were being offered.
68. Korman, "The Holocaust in American Historical Writing," 270.
69. Nik Cohn, "Finally, the Full Force of the Who," *New York Times*, March 8, 1970, M2.

Chapter 2 Holocaust Education in New York City

1. Jick, "The Holocaust: Its Use and Abuse," 311.
2. Elie Wiesel, "Survivors' Children," *New York Times*, November 16, 1975, Section 11, 36.
3. "Genocide Charged by 3 Ulster M.P.s," *New York Times*, August 18, 1969, 2; "Genocide is Charged by Biafran Unit Here," *New York Times*, January 20, 1970, 3; Michael T. Kaufman, "2 Sudanese Rebels, Charging Genocide Seek Help at U.N.," *New York Times*, January 5, 1971, 6. See also Fox Butterfield, "Mujub Resentful of Nixon's Policy," January 15, 1972, 7; *New York Times*, September 27, 1970, 15; *New York Times*, July 12, 1970.
4. James Naughton, "Nixon Urges Senate to Ratify Genocide Past, Stalled Since '50," *New York Times*, February 20, 1970; see *New York Times*, February 24, 1970, 8; June 26, 1970, 16; August 2, 1970, 45; November 26, 1969, 10; and January 10, 1970, 16.
5. Post, *The Holocaust: A Case Study in Genocide*, Foreword. Throughout the text I will be using "Holocaust" and "genocide" education more or less as synonyms to reflect the way the terms were being used in the 1970s and 1980s. By the 1990s and 2000s, unfortunately with further genocides in Rwanda and Darfur, "genocide education" and "Holocaust education" began to be perceived as related, but distinct areas.
6. Ibid., 2.
7. Ibid., Foreword.

8. Ibid., 5.
9. Ibid., 1.
10. Ibid., 2, 9, 10.
11. Roskies, *Teaching the Holocaust to Children*, 37.
12. "Mandatory Study of Nazi Holocaust," *New York Times,* June 23, 1974, 11.
13. Israel Shanker, "Awesome Reliving of Auschwitz Unfolds at St. John's," *New York Times*, June 4, 1974, A39; "Scholars at Holocaust Conference Here Seek Answers to the Unanswerable," *New York Times,* March 4, 1975, A13; "Holocaust Parley Has Few Answers," March 6, 1975, A9.
14. William Styron, "Auschwitz's Message," *New York Times,* June 25, 1974, A37.
15. Littell, "Breaking the Silence," 202.
16. James Clarity, "Philadelphia Schools to Require a Course on Nazi Holocaust," *New York Times*, September 18, 1977, A23.
17. Board of Education of the City of New York, Division of Curriculum and Instruction, *The Holocaust: A Study in Genocide,* iii, xiv. The version of the curriculum cited earlier was published in 1979. This version was probably revised. I was unable to obtain the original 1977 curriculum that the contemporary critics would have seen, because that version was never published. The assertions I present about the original curriculum were corroborated by a description found in a *New York Times* article by Ari Goldman (October 8, 1977). Since I could not track down Albert Post for an interview, the only source I have is the 1979 version of the curriculum. I suspect only minor changes were made. For these reasons, I am hesitant to cite directly the 1979 version of the curriculum, but in my judgment, the few quotations I did use were likely found in both versions.
18. Ari L. Goldman, "Mixed Reaction on Holocaust Study," *New York Times,* October 8, 1977, A36.
19. See Douglas E. Kneeland, "German-Americans Grow Uneasy," *New York Times,* June 24, 1978, A6.
20. Paul Ronald, letter to the editor, *New York Times,* October 15, 1977, A22; Isle Hoffman and Howard Marcus, letters to the editor, *New York Times,* October 18, 1977, A36.
21. See Lipstadt, *Denying the Holocaust*, chapter 7.
22. George Pape, letter to the editor, *New York Times*, November 21, 1977, A36.
23. Yehuda Bauer, letter to the editor, *New York Times*, October 25, 1977, A38.
24. "Teaching the Holocaust," *New York Times,* November 9, 1977, A25.
25. Walter J. Fellenz, letter to the editor, *New York Times*, December 22, 1977, A16.
26. Philip J. Reiss, letter to the editor, *New York Times,* November 23, 1977.
27. Eileen O'Connor, letter to the editor, *New York Times,* November 23, 1977, A18; Nathan Belth, letter to the editor, *New York Times,* November 30, 1977, A24. Similar complaints erupted in the city of Philadelphia, which first introduced the Holocaust in its schools on a limited basis in 1976 and required it as part of a world history course in dozens of its senior and junior and high schools. James F. Clarity, "Philadelphia Schools to Require a Course on Nazi Holocaust," *New York Times,* September 18, 1977, A23.

28. Novick, *The Holocaust in American Life*, 196. Novick feels that the exceptionalist position put forth by certain Jewish leaders has characterized Jews as "intent on permanent possession of the gold medal in the Victimization Olympics."
29. *The Holocaust: Case Study in Genocide,* iii, xvi, xvii. xx.
30. "A Mandatory Study of Nazi Holocaust Is Urged in Schools," 11.

CHAPTER 3 AFFECTIVE REVOLUTION AND HOLOCAUST EDUCATION

1. Marker and Mehlinger, "Social Studies," 830–51.
2. Evans, *The Social Studies Wars*, 134–35; Hertzberg, *Social Studies Reform*, 121–26.
3. *Social Education* 33 (April 1969).
4. "Social Studies Curriculum Guidelines," *Social Education* 36 (December 1971): 853.
5. Ibid., 860. The rapid shift to ethnic studies and relevance can be seen in the titles of the NCSS yearbooks: *Teaching Ethnic Studies* (1973), *Teaching American History: The Quest for Relevance* (1974), and *Controversial Issues in the Social Studies: A Contemporary Perspective* (1975).
6. Oliver and Shaver, *Teaching Public Issues*, 33.
7. Ibid., 9, 35.
8. See Bohan and Feinberg, "Leader–Writers."
9. Kohlberg and Mayer, "Development as the Aim of Education," 457.
10. Ibid., 455.
11. Ibid.
12. See Harlow, Cummings, and Aberasturi, "Karl Popper and Jean Piaget," 41–48.
13. Ibid., 457.
14. Kohlberg, "Stage and Sequence," 376, 414.
15. Ibid., 393.
16. Kohlberg, "The Cognitive Developmental Approach," 185–86.
17. Ibid., 189.
18. Simon, Howe, and Kirschenbaum, *Values Clarification*, 19; Kohlberg, "The Cognitive Developmental Approach," 185.
19. For more examples of educators' interest in applying affective learning (though not necessarily in the social studies), see Krathwohl, Bloom, and Masia, *Taxonomy of Educational Objectives*; Raths, Harmin, and Simon, *Values and Teaching*; Weinstein and Fantini, *Toward Humanistic Education*; and Thayer, ed., *Affective Education*.
20. See Schlesinger, *The Disuniting of America*.
21. In addition to Oliver and Shaver's reference to the Holocaust (cited earlier), Kohlberg referred to the event in a section called "Scoring of moral judgments of Eichmann for developmental stages," in "The Cognitive-Developmental

Approach to Socialization," 383. As we shall see, the exercise of evaluating historical actors using Kohlberg's moral framework was repeated by the curricula designers and their students. Historical context was apparently insignificant.

22. Raymond Zwerin, phone interview with author, January 31, 2003.
23. Ibid.; Irving Spiegel, "Jews Urged to Teach Youth About the Nazi Crimes," *New York Times,* December 2, 1974, A14; "Jewish Body Urges Holocaust Studies," *New York Times,* June 16, 1975, A6.
24. Zwerin, interview with author.
25. Alternatives in Religious Education Publishing Company, "About A.R.E. Publishing," http: arepublish.com/about.htm (viewed October 29, 2003).
26. Zwerin, Marcus, and Kramish, *The Holocaust*, 1, 2, 4.
27. Ibid., 7–12.
28. Zwerin, Marcus, and Kramish, *Gestapo*. I arranged to observe the *Gestapo* game being played by several volunteer teachers (elementary and secondary) and, in this case, all the participants survived. The players had a basic understanding of the chronology of the Holocaust and quickly realized that they should risk their life markers early on in the game. Anticipating the concentration camps to come, they knew that their lives were at minimal risk early in the game, or at least they knew that the situation would get much worse as the game progressed. Of course, the actual victims of the Holocaust did not have the benefit of this knowledge.
29. Chartock, *An Evaluative Study*, 79.
30. Freedman, "Why Teach About the Holocaust," 263.
31. *Social Education* 42 (April 1978).
32. Chartock, *An Evaluative Study*, 99–100.
33. Ibid., 92, 94, 99; Chartock, "Teaching About the Holocaust," 37.
34. Chartock, *An Evaluative Study*, 102.
35. Chartock, "A Holocaust Unit," 278.
36. Kohlberg, "The Cognitive Developmental Approach."
37. Chartock, "A Holocaust Unit," 280–81.
38. Ibid., 282.
39. Richard Flaim, phone interview, December 19, 2003.
40. Ibid.
41. Rosenblum, "State Backs Holocaust Course," 4.
42. Ibid.
43. Murray Schumach, "Students at Teaneck Agonize Over the Holocaust," *New York Times*, June 12, 1976, A25.
44. Richard Flaim phone interview; Flaim and Reynolds, *The Holocaust and Genocide*, ix.
45. New Jersey Department of Education, New Jersey Commission on Holocaust Education, "Holocaust/Genocide Education, New Jersey 1973–2000," www.state.nj.us/njed/holocaust/about2.htm (accessed November 27, 2002).
46. See Joan Verdon, "Genocide Text Altered to Cover Killings of Non-Jews," *The Record* (Bergen, NJ), November 15, 1985, B01.
47. Flaim and Reynolds, *The Holocaust and Genocide*, iii–iv.

48. Stern Strom, "A Work in Progress," 76.
49. A review of the unit can be found in Roskies, *Teaching the Holocaust to Children*, 25–26. Roskies criticized their approach to the Holocaust as a story that has "to be told to the extent to which it is relevant to me and my experience."
50. Fred M. Newmann with the assistance of Donald W. Oliver, *Clarifying Public Controversy*, 32, 340.
51. Interview with William Parsons, October 27, 2003.
52. Stern Strom, "A Work in Progress."
53. Ibid.
54. Ibid., 91; Stern Strom and Parsons, *Facing History*, 2.
55. A subject search for the word "holocaust" in *New York Times* or *Washington Post* prior to 1960 will demonstrate this point. For more on the uses of the word "holocaust" see Jick, "The Holocaust: Its Use and Abuse," 303–18.
56. Stern Strom and Parsons, *Facing History*, 3, 13.
57. Ibid., 14, 393; Glynn, Bock, and Cohen, American Youth and The Holocaust, 16–23.
58. Ibid., 84–86, 106–7.

CHAPTER 4 WATCHING AND DEFINING THE HOLOCAUST

1. Myra MacPherson and Rob Warden, "The Survivors and the Neo-Nazis: The Impact of the Holocaust in Skokie and at Nazi Party Headquarters," *The Washington Post,* April 20, 1978, B1; George Vecsey, "Christians to Wear Star of David in Support of Jews," *New York Times,* April 13, 1978, A20; "Illinois to Mark Holocaust," *New York Times,* April 12, 1978, A11.
2. *New York Times,* April 21, 1978, B4.
3. Witt, *A Humanities Approach*, 1.
4. Richard Flaim, interview with the author, December 19, 2003.
5. Paul Hodge, "Holocaust Panel Urges Memorial Museum in D.C.," *Washington Post,* September 27, 1979, A1.
6. Richard F. Shepard, "Ethnic Leaders React to the Impact of the Holocaust," *New York Times,* April 16, 1978, A60.
7. Douglas E. Kneeland, "Interest in Holocaust Study Rising," *New York Times,* April 22, 1978, B4.
8. "Holocaust Study Aids and Spin-offs," *New York Times,* April 14, 1978, C26.
9. *The Record: A Holocaust History 1933–1945* (New York: Anti-Defamation League of B'nai B'rith, 1978).
10. "Area Schools Get Study Guides for Holocaust Series," *Washington Post,* September 7, 1979, C15.
11. "Holocaust Study Aids and Spin-offs," C26.
12. Richard F. Shepard, "Ethnic Leaders React to the Impact of 'Holocaust,'" *New York Times,* April 16, 1978, A60; William Safire, "Silence is Guilt," *New York Times,* April 24, 1978, A23.

13. Elie Wiesel, "Telling the War," *New York Times*, November 5, 1972, Section 7, 3.
14. Elie Wiesel, "Trivializing the Holocaust: Semi-Fact and Semi-Fiction," *New York Times,* April 16, 1978, B1.
15. Kolbert, *The Worlds of Elie Wiesel*, 22.
16. Wiesel, *All Rivers Run to the Sea*, 45.
17. Quoted in Estess, *Elie Wiesel*, 115.
18. Sarah Layden, "Silence Worries Nobel Winner Author Elie Wiesel Tells Freshmen that 'You Must be on the Side of the Victim,'" *The Post Standard* (Syracuse, NY), September 28, 2000, B5.
19. Sidney Bolkosky, "The Search for Meaning," http: www.personal,umd. umich.edu/~sbolkosk/meaning.htm (accessed October 12, 2003).
20. Franciosi, ed. *Elie Wiesel*, 23.
21. Abrahansom, *Against Silence*, 158.
22. Wiesel, "Then and Now," 270.
23. Ibid., 269.
24. Ibid., 271.
25. Ibid.
26. Ibid.
27. Ibid., 270.
28. Douglas Kneedland, "German-Americans Grow Uneasy," *New York Times*, June 24, 1978; Israel Shenker, "Holocaust Survivors Remember," *New York Times*, April 20, 1978; Paul Dobin, letter to the editor, *New York Times,* May 1, 1978; Joseph W. Eaton, letter to the editor, *New York Times,* April 19, 1978, Richard Shepard, "TV' Story of Nazi Terror Brings High Ratings and Varied Opinions," *New York Times,* April 18, 1978; "Images of Holocaust," *New York Times*, April 16, 1978; William Safire, "Silence is Guilt," *New York Times*, April 24, 1978.
29. For examples of those who conflate the elements of Holocaust uniqueness, see Littell, "Breaking the Silence," 195–212; Ben-Bassat, "Holocaust Awareness and Education in the United States," 403–23; Totten "Holocaust Education in the United States," 305–12.
30. Paula Hyman, "The New Debate on the Holocaust," *New York Times,* September 14, 1980, F65.
31. Mintz, *Popular Culture*, 38–39.
32. Quoted in Friedlander, "Towards a Methodology of Teaching," 524.
33. Ibid., 524. For more on Fackenheim's views, see Fackenheim, "Concerning Authentic Responses to the Holocaust," 68–87.
34. Ibid., B29.
35. For a critique of Goldhagen, see the "Afterword" in Christopher Browning's study *Ordinary Men*. See also Shandley, ed., *Unwilling Germans?*
36. Young, "Towards a Received History of the Holocaust," 36–37.
37. For a discussion of this debate see Friedlander, ed., *Probing the Limits of Representation.*
38. See Magurshak, "The 'Incomprehensibility' of the Holocaust," and Bauer, "Against Mystification," in *The Nazi Holocaust,* 88–117.

39. Lispstadt, *Denying the Holocaust*, 213.
40. Katz, *The Holocaust in Historical Context*. For a critique of Katz's thesis, see chapter five in Lang, *The Future of the Holocaust*.
41. Fackenheim, "The Holocaust and Philosophy," 506.
42. Ibid.
43. Margalit and Motzkin, "The Uniqueness of the Holocaust," 83.
44. Yehuda Bauer, "Extreme and Unique Holocaust," *New York Times*, October 25, 1977, A38.
45. Novick, *The Holocaust in American Life*, 196.
46. Clendinnen, *Reading the Holocaust*, 4.
47. Ibid., 183.
48. Dawidowicz, *The Holocaust and the Historians*, 13.
49. For a small sample of the literature on the Americanization of the Holocaust, see Flanzbaum, ed., *Americanization of the Holocaust*; Rosenfeld, "The Americanization of the Holocaust," 35–40; and Linenthal, "The Boundaries of Memory," 406–33.
50. In addition, the curricula associated with "values clarification" came under attack. See Bennett, "What Value is Values Education?" 31–32.
51. Hertzberg, *Social Studies Reform*, 169.
52. Evans, *The Social Studies Wars*, 2004, 149–53; Ravitch, *Left Back*, 408–20.
53. National Commission on Excellence in Education, *A Nation At Risk: The Imperative For Educational Reform* (Washington D.C.: U.S. Government Printing Office, 1983), 5.
54. Ravitch, *Left Back*, 411–14.
55. Quoted in McClellan, *Moral Education in America*, 92.
56. Zwerin, interview with author.
57. This research originally appeared in Schweber, *Teaching History, Teaching Morality*. It was later published as "Simulating Survival," *Curriculum Inquiry* 33:2 (2003) and *Making Sense of the Holocaust: Lessons from Classroom Practice* (New York: Teachers College Press, 2003).
58. Roskies, *Teaching the Holocaust to Children*, 16. Toppin, *After the War and the Election of 2020*.
59. Sheramy, *Defining Lessons*, 86.
60. Paul Hyman, "New Debate on the Holocaust," *New York Times*, September 14, 1980, Section 6, 65.
61. Ibid.
62. Zwerin, interview with author.
63. This exercise was documented in the short film *Eye of the Storm*.
64. Zwerin, interview with author.
65. Novick, *The Holocaust in American Life*, 184–85.
66. Joan Verdon, "Genocide Text Altered To Cover Killings of Non-Jews," *The Record*, November 15, 1985, B01.
67. Ibid.
68. This threat was allegedly related to Flaim through representatives from both the ADL and the New Jersey Board of Education.
69. Ibid.; Richard Flaim, phone interview with author.

70. Ibid.
71. Ed Vulliamy, "Holocaust Project Funds: 'Eliminated' by Ideology," *Washington Post,* October 4, 1988, A17.
72. "Fund Denial for Holocaust Course Is Upheld," *Washington Post,* January 5, 1989, A19.
73. Bill McAllister, "Education Dept. Clears Holocaust Study Grant," *Washington Post*, September 27, 1989, A15.
74. Dawidowicz, "How They Teach the Holocaust," 65–66, 73–74.
75. Ibid., 79.
76. Stephen Labaton, "Hello, Goodbye: The Short History of a House Historian," *New York Times,* January 15, 1995, Section 4, 2.
77. Lipstadt, "Not Facing History," 26–29. For a more in-depth (and biased) depiction of the public debate over *Facing History,* see Fine, *Habits of Mind.*
78. Christina Jeffrey, "Christina Jeffrey Responds," *Washington Post,* January 24, 1995, A16.
79. Glynn, Bock, and Cohen, *American Youth and the Holocaust.*
80. Amitai Etzioni, "Who's To Say What's Right or Wrong?" *Washington Post,* April 2, 1995, R14.
81. Jeff Jacoby, "Facing History, What It Really Teaches," *Boston Globe,* January 17, 1995, 9.
82. Glynn, Bock, and Cohen, *American Youth and the Holocaust,* 130.

Chapter 5 Holocaustomania

1. William Honan, "Holocaust Teaching Gaining a Niche, but Method is Disputed," *New York Times*, April 12, 1995, B11.
2. Paul Hyman, "New Debate on the Holocaust," *New York Times,* September 14, 1980, F65; Kenneth L. Woodward with Eloise Salholz, "Debate Over the Holocaust," *Newsweek,* March 10, 1980, 97; Lawrence Feinburg, "The New Impact of the Holocaust," *Washington Post,* October 10, 1979, C1.
3. One exception was a 1978 article in the *Chronicle of Higher Education* by Ellen Coughlin, which pointed out in its title that "On University Campuses, Interest in the Holocaust Started Long Ago." May 1, 1978, 1, 8.
4. Russell Chandler, "1st Center for Holocaust Studies To Be Established in Los Angles," *Washington Post,* August 26, 1977, C8. Similar centers were established at Yeshiva University in Brooklyn (1975), Temple University (1975), and Brandeis University (1980).
5. Alter, "Defamations of the Holocaust," 49.
6. The first estimate, ninety-three, was by Robert Alter (ibid.) and the second, seven hundred, was by Franklin Littell in Coughlin, "On University Campuses, Interest in the Holocaust Started Long Ago," 8.
7. Friedlander, *On the Holocaust.*
8. Fred Hechinger, "About Education: Educators Seek To Teach Context Of the Holocaust," *New York Times,* May 15, 1979, C5.
9. Friedlander, "Toward a Methodology," 520–22.

10. Ibid., 525.
11. Ibid., 531–32.
12. Ibid., 522, 526, 532–33.
13. Quoted in Ira Rosenblum, "State Backs Holocaust Course," *New York Times*, February 17, 1985, Section 11NJ, 4.
14. Quoted in Pate, "The Holocaust in American Textbooks," 245.
15. In many cases these textbook critiques were intended for a general readership. Two of the more popular studies were Fitzgerald, *America Revised* and Loewen, *Lies My Teacher Told Me*. A more recent example is Ravitch, *The Language Police*.
16. Sheramy, *Defining Lessons*, 12.
17. Keith, "Politics of Textbook Selection," 9.
18. Apple, "The Political Economy of Text Publishing," 309.
19. Keith, "Politics of Textbook Selection."
20. Drummond Ayres, "School Critics Press Drive for Old Values," *New York Times*, July 25, 1975, A7.
21. Edwin Barber, "Textbook: Much Labor and Risk," *New York Times*, March 23, 1975, Section 4, 16.
22. Krefetz, "Nazism: The Textbook Treatment," 5–7.
23. Elkin, "Minorities in Textbooks," 504; "Bias Charged in Book Rejection," *New York Times*, November 10, 1974, A53; Loewen, *Lies My Teacher Told Me*, 281.
24. *Social Education* 33 (April 1969). See also Elkin, "Minorities in Textbooks"; and "Survey of Textbooks Detects Less Bias against Blacks," *New York Times*, March 28, 1973, A13.
25. Keith, "Politics of Textbook Selection," 31.
26. Hilberg, "Developments in the Historiography of the Holocaust," 21.
27. Crowe, "The Holocaust, Historiography and History," 24–61; Hilberg, "Developments in the Historiography of the Holocaust"; Dawidowicz, *The Holocaust and the Historians*; United States Holocaust Memorial Museum, *Teaching About the Holocaust*.

    By the mid 1970s, the literature on the Holocaust was considerable. I chose to highlight the work of Dawidowicz and Bauer for two reasons. First, these two historians appear throughout the narrative of this book. Second, because these are two works that are commonly cited by Holocaust educators as being "influential" on their thought and teaching. For a thorough review of the available literature in 1973, see Friedlander, *On the Holocaust*.
28. Koka, "German History before Hitler," 3–16; Maier, *The Unmasterable Past*, 113–14.
29. Rathenow, "Teaching the Holocaust in Germany," 63–76.
30. Maier, *The Unmasterable Past*, 84.
31. Flaim, phone interview with author, December 19, 2003.
32. Despite this, critiques of textbook coverage of the Holocaust continued into the 1990s. For example, see Kanter, "Forgetting to Remember."
33. Ravitch and Finn, *What Do Our 17-Year Olds Know?*, 49, 61.

34. Joseph Berger, "Once Rarely Explored, the Holocaust Gains Momentum as a School Topic," *New York Times,* October 3, 1988, A16.
35. An excellent resource for tracking state Holocaust and genocide legislation can be found on the USHMM's webpage at http://www.ushmm.org/education/foreducators/states/index.php?state=WY.
36. Brabham, "Holocaust Education," 139–42.
37. Alex Patricia, "Teaching the Holocaust; Genocide Curriculum, as Mandated by the State, Gradually Takes Shape," *The Record* (Bergen, NJ), August 30, 1994, B01.
38. Elizabeth Llorente, "Genocide Study Guides Delayed Until New School Year; Still Being Reviewed, Panel Says" *The Record,* April 26, 1996, A08; "Irish Group in Genocide Study Fight; Wants Famine in Curriculum," April 30, 1996, A04.
39. Sarah Metzgar, "Legislator Wants State to Teach Famine History," *The Times Union* (Albany, NY), February 29, 1996, A1.
40. Cavaiani, "Yet Another Unfair Demand on Schools," *The Record* (Bergen, NJ), September 15, 1998, L08.
41. *Final Report to the Governor from Ohio Council on Holocaust Education* (December 1987), Appendix A Executive Order 86-24, Creating The Ohio Council on Holocaust Education, 10.
42. It is very likely that these teachers believed in both forms of "universalizing" the Holocaust, but the hastily designed survey was designed to allow only one answer. It also would have been helpful if the survey asked from where the teachers had received their curricula and information. *Final Report to the Governor from Ohio Council on Holocaust Education,* 34–35.
43. Danks, "An Unlikely Journey," 90.
44. Rabinsky, "A Journey Through Memory," 123–35.
45. Danks, "An Unlikely Journey," 89.
46. Rabinsky, "A Journey Through Memory," 43.
47. Rabinsky and Danks, *The Holocaust: Prejudice Unleashed*, 4.
48. Frampton, *A Descriptive Study To Ascertain Curriculum Guidelines*, 96, 99.
49. Flaim, phone interview with author, December 19, 2003.
50. *Final Report to the Governor from Ohio Council on Holocaust Education,* 15, 29.
51. Rabinsky, "A Journey Through Memory," 36.

CHAPTER 6 CRITIQUING HOLOCAUST EDUCATION

1. Quoted in Mintz, *Popular Culture,* 145. This entire section draws upon chapter three of Mintz's book.
2. Goldstein, "Teaching Schindler's List," 362–64.
3. Goldberg, "The Holocaust and Education," 318.
4. Short and Reed, *Issues in Holocaust Education*, 16.

5. Ibid.,12–25; Hector, "Teaching the Holocaust in England," 105–16.
6. Quotation taken from Eugene Kiely, "Prejudice, Health are Governor's Concerns: Whitman Putting Tolerance Program on New Jersey List," *The Record* (Bergen, NJ), July 19, 1994, A03.
7. Wiesel, *And the Sea is Never Full,* 181–82, 204.
8. Ibid., 186.
9. "Address by President Jimmy Carter," Appendix C, *Report to the President: President's Commission on the Holocaust,* 26.
10. Linenthal, "The Boundaries of Memory," 431.
11. Seymour Bolten, quoted in Linenthal, *Preserving Memory,* 43.
12. Quoted in Flunzbaum, "Introduction: The Americanization of the Holocaust," in *The Americanization of the Holocaust,* 5.
13. Linenthal, *Preserving Memory,* 133, 129–30; Wiesel, *And the Sea is Never Full,* 246.
14. It is no surprise that Wiesel was not pleased with the final exhibit. He expressed these views in his critical response to the USHMM's exhibit. He wrote, "the museum by trying to say everything, does not say enough... Upon leaving, the visitor will be able to say 'Now I know everything; I understand.' Later he or she will say: 'I was there.' I had a different vision of the museum. I should have liked the visitor to leave saying: 'Now I know how little I know'... In trying to show everything, you conceal the essential... In this case the saying 'less is more' is apt... in truth I would have preferred a more sober, more humble edifice, one that would suggest the unspoken, the silence, the secret."
15. *Report to the President: President's Commission on the Holocaust,* 13.
16. William Parsons, interview with author, October 27, 2003.
17. William Parsons gave a copy of the "Mission Statement" to me. It also appears throughout the USHMM literature.
18. Parsons, interview with author.
19. In addition to Parsons and Totten, Steven Feinberg, William Fernekes, Sybil Milton, and Sara Bloomfield also contributed to the writing of the USHMM Guidelines.
20. Totten, "Why?," 178–79, 180–90.
21. Samuel Totten, interview with author (electronic mail), November 16, 2003.
22. Totten, "Why?," 209. Here Totten was referring to Benjamin S. Bloom's taxonomy of educational objectives, which established a hierarchy of instructional outcomes: knowledge, comprehension, application, analysis, synthesis, and evaluation. See Bloom et al., eds., *Taxonomy of Educational Objectives.*
23. Totten, interview with author.
24. Ibid.
25. Parsons and Totten, *Guidelines,* 3.
26. Joseph Berger, "Once Rarely Explored, the Holocaust Gains Momentum as a School Topic," *New York Times,* October 3, 1988, 16.
27. Parsons and Totten, *Guidelines,* 6.

28. See Ravitch and Finn, *What Do Our 17-year Olds Know?*; Bradley Commission on History in Schools, *Building a History Curriculum: Guidelines for Teaching History in Schools* (Washington, D.C., 1998); Nash, Crabtree, and Dunn, *History on Trial*; Wineburg, *Historical Thinking*.
29. Parsons and Totten, *Guidelines*, 8.
30. For the best discussion of the concept of historical empathy, see chapters eleven and twelve in Barton and Levstik, *Teaching History for the Common Good*.
31. Riley and Totten, "Understanding Matters," 559.
32. See Totten and Riley, "Authentic Pedagogy and the Holocaust," 120–41; Totten, "The Start is as Important as the Finish," 70–76; "A Holocaust Curriculum," 148–66. Most of his published work has been collected in Totten, *Holocaust Education*.
33. Totten, interview with author.
34. Totten, "Diminishing the Complexity," 165, 170.
35. Totten, *Holocaust Education,* 167. Totten was critiquing Harriet Sepinwall, "Incorporating Holocaust Education," 5–8.
36. Totten, interview with author.
37. Ibid.
38. Riley, "The Holocaust and Historical Empathy," 158.
39. Ibid., 150, 153.
40. Shawn, "Current Issues in Holocaust Education," 16, 18.
41. Stephen W. Wylen, "Holocaust Education Should Touch a Raw Nerve; Lessons Must Look Directly at Human Evil," *The Record,* November 9, 2000, H6.
42. Totten, *Holocaust Education, 79.*
43. Sidney Bolkosky, review of Samuel Totten, *Holocaust Education* in *Holocaust and Genocide Studies*, 18 (Fall 2004): 309. Other essays offering advice on how to teach the event include Glanz, "Ten Suggestions for Teaching the Holocaust," 547–65 and Schweber, "Holocaust Fatigue in Teaching Today," 44–49. Glanz suggests that teachers encourage a hands-on and minds-on approach to learning, employ a K-W-L strategy (what I know, what I want to know, what I learned), facilitate many class discussions, supported by appropriate reading assignments and reinforced by reflective journal writing, use videos liberally and intelligently, emphasize the rich cultural heritage of the Jewish community prior to the Holocaust, invite survivors as guest speakers, visit the USHMM, provide a firm knowledge base of historical events, explore websites on the Internet, and to "never forget why you are a teacher." Glanz asserts that "Holocaust study provides a forum to sensitize students to human suffering and oppression as well as to encourage an 'ethic of caring' for all people" (p. 562). Drawing upon her numerous ethnographic studies of Holocaust teachers (see chapter seven), Schweber suggests that educators teach about the history of anti-Semitism, provide students with a range of explanations for perpetrators' behaviors, use popularizations and current uses of the Holocaust as teachable texts, and "know

here you draw the boundaries between over generalizing and over specifying" (p. 49).

44. Ravitch, "Who Prepares Our History Teachers?" 495–503.

Chapter 7 Out of the Discourse, Into the Classroom

1. Morse, *While Six Million Dies*; Feingold, *The Politics of Rescue*; Penkower, *The Jews Were Expendable*; and Gilbert, *Auschwitz and the Allies.*
2. Wyman, *The Abandonment of the Jews.*
3. Lipstadt, *Beyond Belief.*
4. Novick, *The Holocaust in American Life* and Dawidowicz, "Could America Have Rescued the Jews?" 157–78.
5. Shandler, *While America Watches*; Flunzbaum, ed., *Americanization of the Holocaust*; Loshitzky, ed., *Spielberg's Holocaust*; Avisar *Screening the Holocaust*; Doneson, *The Holocaust in American Film*; and Insdorf, *Indelible Shadows.*
6. For a description of the political, cultural, and logistical struggle to design and build the United States Holocaust Memorial Museum, see Linenthal, *Preserving Memory.* For a concise version of this research, see Linenthal, "The Boundaries of Memory," 406–33.
7. Cole, *Selling the Holocaust*, 150.
8. Ibid., 260–72.
9. Young, "America's Holocaust," 73.
10. Ibid., 80.
11. See Young, *The Texture of Memory* and Young, ed., *The Art of Memory.*
12. Young, *The Texture of Memory,* 322.
13. See Lagermann, *An Elusive Science.*
14. Epstein, "Holocaust Education in the United States," 1171, 1172.
15. Locke, "The Holocaust and the American University," 1188–93.
16. Haynes, "Holocaust Education at American Colleges and Universities," 294–95, 302.
17. Kanter, "Forgetting to Remember," 43.
18. Ibid., 21. Political Science texts faired slightly better; coverage ranged from zero to sixty-seven lines.
19. Ibid., 41.
20. Elizabeth Llorente, "Holocaust Taught at Nearly all Districts, Survey Finds," *The Record* (Bergen, NJ), April 3, 1998, A03.
21. Donvito, *A Descriptive Study*, 196.
22. Ellison, *From One Generation to the Next*, 45.
23. Holt, *Implementation of Indiana's Resolution.*
24. Donoho, *The Arkansas Holocaust Education*, 117, 140.
25. Mitchell, *Methods of Teaching the Holocaust to Secondary Students*, 133.
26. Ibid., 140.

27. Donoho, *The Arkansas Holocaust Education,* 155.
28. Quoted in William H. Honan, "Holocaust Teaching Gaining a Niche, but Method is Disputed," *New York Times*, April 12, 1995, B11.
29. Schweber "Simulating Survival," 141, 143.
30. Ibid., 148.
31. Ibid., 156.
32. Ibid., 183.
33. Ben-Peretz, "Identifying with Horror," 191, 192.
34. Schweber, "Rejoinder to Miriam Ben-Peretz," 201.
35. Schweber, "'Breaking Down Barriers,'" 28.
36. Schweber, *Making Sense of the Holocaust*, 17.
37. Ibid., 57–58.
38. Ibid., 140.
39. Ibid., 152.
40. Fine, "Collaborative Innovations," 788.
41. Ibid., 776.
42. Ibid., 785.
43. Ibid., 786.
44. Fine, "'You Can't Just Say.'" Accessed May 21, 2007 from www.edreview.org/harvard93/wi93/w93fin.html, 7.
45. Ibid., 8.
46. Ibid., 11.
47. Ibid., 12.
48. Schweber and Irwin, "'Especially Special,'" 1700.
49. Ibid., 1707.
50. Ibid., 1708.
51. Schweber, "Holocaust Education at Lubavitch Girls Yeshiva," 24.
52. Ibid., 27, 32.
53. Schweber, "What Happened to Their Pets?" 12.
54. Ibid., 37, 38.
55. Ibid., 57, 58.
56. Barton and Levstik, "Back When God Was Around and Everything," 419–54.
57. Friedlander, "Towards a Methodology," 519–42.
58. Wegner, "What Lessons Are There from the Holocaust," 176.
59. Lieberman, "Facing History and Ourselves," 36–41.
60. Bardige, "Facing History and Ourselves," 42–48; Brabeck, Kenny, Stryker, Tollefson, and Stern Strom, "Human Rights Education," 333–47; Schultz, Barr, and Selman, "The Value of a Developmental Approach," 3–27.
61. Boix-Manilla, "Historical Understanding," 410.
62. Ibid., 407, 408.
63. Ibid., 406, 412.
64. Spector, "God on the Gallows," 33.
65. Ibid., 36.
66. Ibid., 43.
67. In fact, in a 2002 study, David Hays Lindquist interviewed six teaching fellows at the USHMM about their approaches to teaching the Holocaust,

and compared their answers with the literature on teaching the event (most of it written by Sam Totten). Unsurprisingly, there was almost universal agreement between the two. See Lindquist, *Towards a Pedagogy*.

Chapter 8 Teaching the Holocaust and the Aims of Secondary Education

1. For other examples of how minority groups have exerted pressure on the curriculum through the years, see Moreau, *Schoolbook Nation*; Zimmerman, *Whose America?*; Tyack, *Seeking Common Ground*. These works, however, view the curricular disputes purely in cultural, not pedagogical terms. In many ways, my narrative specifically aimed to move beyond the limits of these interpretations.
2. Smith, *Holocaust Denial*, 46.
3. One may argue that I am drawing too strong a distinction between culture and pedagogy. After all, John Dewey and his contemporaries designed their progressive educational ideas specifically for a democratic American society. Therefore, applying their progressive pedagogy to any content is, in essence, "Americanizing" it. This is a valid objection, but one that ultimately has to do with how one chooses to define the line between culture and education. I have made my distinction based on the motivations of the teachers themselves. They did not approach the curriculum in a certain way because they felt pressured to make the Holocaust politically acceptable to all Americans, to avoid offending certain groups, or to provide their students with a happy ending—accommodations usually associated with Americanizing the Holocaust for films, memorials, and museums. Instead they drew upon pedagogical research and their own classroom experience. For the most part, this cut across religious and cultural lines. Except in the case of religious fundamentalism, whether a teacher was a Jew or gentile had a marginal affect on how they chose to approach the event.
4. On the founding of the social studies, see Fallace, "Did the Social Studies Really Replace History?"
5. Dewey, *Experience and Education*, 17.
6. Hertzberg, *Social Studies Reform*, 65–72. See also "Introduction" in Wineburg, *Historical Thinking*.
7. Bestor, *Educational Wastelands*, 2, 10.
8. Hirsch, "Cultural Literacy," 164
9. Hirsch, *Cultural Literacy*.
10. Quoted in Osborne, "Fred Marrow Fling," 487–48.
11. Quoted in Ravitch, *The Troubled Crusade*, 232.
12. Hertzberg, *Social Studies Reform*, 96–100; Marker and Mehlinger, "Social Studies," 338.
13. Hertzberg, *Social Studies Reform*, 101.

14. Gardner, *The Disciplined Mind*; *The Unschooled Mind*. For Gardner's discussion of the connection between multiple intelligences and disciplinary thinking, see *Multiple Intelligences*.
15. Gardner, *The Disciplined* Mind, 53–54, 157.
16. Ibid., 199.
17. Wineburg, *Historical Thinking and Other Unnatural Acts*, 5.
18. See Gilligan, "In a Different Voice," 481–517. As Gilligan explains in this work, Kohlberg identifies "a strong interpersonal bias in the moral judgments of women, which leads them to be considered as typically at the third of his six-stage developmental sequence...And yet therein lies the paradox, for the very traits that have traditionally defined the 'goodness of woman,' their care for and sensitivity to the needs of others, are those that mark them as deficient in moral development" (p. 484).
19. Kegan, *In Over Our Heads*, 32. I thank my colleague Wendy Atwell-Vasey for pointing me toward Kegan's work. In this text, Kegan defends Kohlberg against Gilligan's accusation mentioned earlier, by suggesting that feminine and masculine styles while indeed different are still related to overall orders of consciousness. "A relational preference, or 'connected' way of knowing," he explains, "does not refer to an inability to differentiate from any epistemological principle, cross-categorical or any other; instead it names a preferred way of relating to that from which one is differentiated" (p. 219).
20. Ibid., 34.
21. Ibid.
22. Kegan, *In Over Our Heads*, 326.
23. Friedlander wrote, "Most [students] assume that all were killing centers and that the methods used at Auschwitz and Treblinka applied also at Dachau and Buchenwald." See Friedlander, "Toward a Methodology of Teaching," 537. Riley wrote, "teachers must help students to cope with factual truths about the Holocaust—Auschwitz was an extermination camp, while Dachau was a concentration camp—and provide opportunities for them to discover these truths through sustained examination of the sources." See Riley, "The Holocaust and Historical Empathy," 140. In response, I would ask, why? Is it really a good use of class time to examine sources to discover the differences between Dachau and Auschwitz? Exactly what sources would outline these differences? Would knowing the difference between the two camps have any long-term effects on the outcomes of a unit on the Holocaust? Friedlander and Riley's critique of the knowledge of students and teachers—in otherwise thoughtful essays—seems to me like intellectual bullying. See chapter six.
24. See Wineburg, "Reading Historical Texts," 495–519. In this study, he outlined the difference in the reading of primary documents between successful high-school students and professional historians.
25. Schweber "Simulating Survival," 204.
26. One element that has been missing from the literature on Holocaust education was, as I outlined in the introduction to this text, American scientists' own role in eugenics and racialist thinking during the first half of the twentieth

century. Perhaps comparing past and present genocides and discriminations is less useful than comparing German racism in the 1930s to that of other countries at that time, including the United States (i.e., Jim Crow South, immigration policy, Japanese internment). This would both make the Holocaust relevant, as well as protect it historical uniqueness.

27. In Schweber's cases of the Christian fundamentalists and Jewish Orthodox teachers Ms. Barrett and Mrs. Glickman, students were shielded from the potential cognitive dissonance of the Holocaust. The Holocaust was deliberately kept at the level of second-order consciousness.
28. To use a metaphor I sometimes use with my preservice teachers to explain historical thinking at different stages, imagine history as a wall of facts and a historian as a painter with a bucket of paint. At first-order consciousness, a student would only see that part of the wall (already painted) that s/he could touch and view. At second-order consciousness, s/he would see the whole wall of facts (already painted), including those parts s/he could not touch. S/he could appreciate that the historian painted the wall, but it would only be painted one color. At third-order consciousness, s/he would appreciate that the wall did not exist independent of the historians having painted it. The facts were not on the wall, but instead in the buckets of historians' paint. The wall would be viewed as multicolored, the product of multiple historians. At, fourth-order consciousness, s/her would appreciate the significance of the colors in each bucket of paint and how the colors limited and directed the act of painting. Finally, at fifth-order consciousness, s/he would appreciate that the buckets, paint, and historians were all arbitrary and designed to exclude those who did not have access to the materials. No color or painter is necessarily better than another.
29. VanSledright, *In Search of America's Past.*
30. See note 21.
31. My suspicion has been confirmed by recent research, which directly targeted the cross-domain stages theory of Piaget in regards to teaching young children history. See Levstik and Pappas, "New Directions for Studying Historical Understanding," 369–85 and Barton and Levstik, "Back When God Was Around," 419–54.
32. Schweber, "Simulating Survival," 158–59.
33. Ibid., 170. There is really not enough to work with here, but this gender discrepancy could possible support Gilligan's critique of Kohlberg's theory. See note 15.
34. Lang, *Act and Idea in the Nazi Genocide*, 25.
35. Ibid., 22.
36. Dewey, "Religion and Morality," 175, 176. I could not locate any writings in which Dewey directly addressed the extermination of the European Jews.
37. Oliver and Shaver, *Teaching Public Issues*, 10.
38. Kohlberg and Mayer, "Development as Aim of Education," 484.
39. Samuel Totten, interview with author, November 15, 2003.
40. Schweber, "Holocaust Education at Lubavitch Girls Yeshiva," 42.
41. Nash, Crabtree, and Dunn. *History on Trial*, 36.

42. See Powell, Farrar, and Cohen, *The Shopping Mall High School*; Goodlad, *A Place Called School*; and Sizer, *Horace's Compromise*.
43. Franklin and McCulloch, "Introduction." On the history of the comprehensive high school see Wraga, *Democracy's High School*.

EPILOGUE: THE FUTURE OF HOLOCAUST EDUCATION

1. See *Discussion and Resource Guide*. The guide includes the contents of the USHMM's "Guidelines for Teaching about the Holocaust" as well as some suggestions for preparing students to watch the documentary.
2. See VanFossen, "'Reading and Math,'" 376–403 and Rock, Heafner, O'Connor, Passe, Oldendorf, Good, and Byrd, "One State Closer to a National Crisis," 455–83.
3. For the effect of standards on the teaching of history, see Grant, *History Lessons*; Grant, ed., *Measuring History*; Yeager and Davis, eds., *Wise Social Studies Teaching*.

Bibliography

Holocaust Curricula and Educational Materials

Anti-Defamation League. *The Record: A Holocaust History 1933–1945*. New York: Anti-Defamation League of B'nai B'rith, 1978.

Board of Education of the City of New York, Division of Curriculum and Instruction. *The Holocaust: A Study in Genocide*. New York: Board of Education of the City of New York, 1979.

Chartock, Roselle. "A Holocaust Unit for Classroom Teachers." *Social Education* 42 (April 1978): 278–85.

Doerr Toppin, Martha. *After the War and the Election of 2020*. Florida Holocaust Resource and Education Center, 1980.

Flaim, Richard and Edwin Reynolds. *The Holocaust and Genocide: A Search for Conscience, A Curriculum Guide*. New York: Anti-Defamation League, 1983.

Goldstein, Phyllis. "Teaching Schindler's List." *Social Education* 59 (October 1995): 362–64.

Post, Albert. *The Holocaust: A Case Study in Genocide: A Teaching Guide*. New York:American Association for Jewish Education, 1973.

Rabinsky, Leatrice and Carol Danks. *The Holocaust: Prejudice Unleashed*. Columbus, OH: Ohio Council on Holocaust Education, 1989.

Spotts, Leon. *Guide to Teachers and Group Leaders: The Story of the Jewish Catastrophe in Europe*. New York: National Curriculum Research Institute, 1967.

Stadtler, Bea. *The Holocaust: A History of Courage and Resistance*. New York: Behrman House, Inc, 1974.

Stern, Jay B. *Learning More about Pathways through Jewish History*. New York: KTAV Publishing, 1976.

Stern Strom, Margot. *Facing History and Ourselves: Holocaust and Human Behavior*. 2nd edition. Brookline, MA: Facing History and Ourselves, 1994.

Stern Strom, Margot and William Parsons. *Facing History and Ourselves: Holocaust and Human Behavior*. Brookline, MA: Facing History and Ourselves, 1982.

United States Holocaust Memorial Museum. *Teaching about the Holocaust: A Resource Book for Educators.* Washington, D.C.: USHMM, 2001.
Zwerin, Raymond, Audrey Friedman Marcus, and Leonard Kramish. *The Holocaust: A Study in Values.* Denver: Alternatives in Religious Education, 1976.
———. *Gestapo: A Learning Experience about the Holocaust.* Denver: Alternatives in Religious Education, 1976.

Primary Sources

Boston Globe
Commentary
Congress Bi-Weekly
Conservative Judaism
Chronicle of Higher Education
Educational Theory
Jewish Education
Modern Judaism
New York Times
The Reconstructionist
The Record (Bergen, NJ)
Religious Education
Social Education
Societas
The Times Union (Albany, NY)
Today's Education
Washington Post
Yad Vashem

Interviews

Richard Flaim, phone interview with author, December 19, 2003.
William Parsons, interview with author, October 27, 2003.
Samuel Totten, electronic interview with author, November 16, 2003
Rabbi Raymond Zwerin, phone interview with author, January 31, 2003.

Secondary Sources

Abrahansom, Irving. *Against Silence: The Voice and Vision of Elie Wiesel*, vol. 1. New York: Holocaust Library, 1985.

Alter, Robert. "Defamations of the Holocaust." *Commentary* 71:2 (February 1981): 49.

Apple, Michael W. "The Political Economy of Text Publishing." *Educational Theory* 34 (Fall 1984): 307–19.

Avisar, Ilan. *Screening the Holocaust: Cinema's Images of the Unimaginable.* Bloomington: Indiana University Press, 1988.

Bardige, Betty. "Facing History and Ourselves: Tracing Development through Analysis of Student Journals." *Moral Education Forum* 6 (1981): 42–48.

Baron, Lawrence. "The Holocaust and American Public Memory, 1945–1960." *Holocaust and Genocide Studies* 17:1 (Spring 2003): 62–88.

Barton, Keith and Linda Levstik. *Teaching History for the Common Good.* Mahwah, NJ: Lawrence Erlbaum Associates, 2004.

———. "Back When God wWas Around and Everything: The Development of Children's Understanding of Historical Time." *American Educational Research Journal* 33 (Summer 1996): 419–54.

Bauer, Yehuda. "Against Mystification: The Holocaust as a Historical Phenomenon." In *The Nazi Holocaust: Perspectives on the Holocaust,* vol. 1. Edited by Michael R. Marrus. Westport: Meckler, 1989: 88–117.

Ben-Bassat, Nurith. "Holocaust Awareness and Education in the United States." *Religious Education* 95 (Fall 2000): 403–23.

Ben-Horin, Meir. "Teaching about the Holocaust." *Reconstructionist* 27 (May 5, 1961): 5–9.

Bennett, Alan D. "Towards a Holocaust Curriculum." *Jewish Education* 43 (Spring 1974): 22–26.

Bestor, Arthur. *Educational Wastelands: The Retreat from Learning in Our Public Schools.* Urbana: University of Illinois Press, 1953.

Bloom, B.S., M.B. Englehart, E.J. Furst, W.H. Hill, and D.R. Krathwohl. eds. *Taxonomy of Educational Objectives: The Classification of Educational Objectives. The Classification of Educational Goals, Handbook I Cognitive Domain.* New York: McCay, 1956.

Blumberg, Herman J. "Some Problems in Teaching the Holocaust." *Reconstructionist* 34 (December 20, 1968): 13–20.

Bohan, Chara Haeussler and Joseph Feinberg. "A Rebellious Jersey Girl: Rachel Davis DuBois, Intercultural Education Pioneer." In *Addressing Social Issues in the Classroom and Beyond: The Pedagogical Efforts of Pioneers in the Field.* Edited by Samuel Totten and Jon Pedersen. Charlotte, NC: Information Age Publishing, 2007: 99–115.

———. "Leader–Writers: The Contributions of Oliver, Fred Newmann, and James Shaver to the Harvard Social Studies Project." In *New Social Studies: People, Projects, Politics, Perspectives.* Edited by Barbara Slater Stern (forthcoming).

Boix-Manilla, Veronica. "Historical Understanding: Beyond the Past and into the Present." In *Knowing, Teaching and Learning History: National and International Perspectives.* Edited by Peter Sterns, Peter Seixas, and Sam Wineburg. New York: New York University Press, 2000: 390–418.

Brabeck, Mary, Maureen Kenny, Sonia Stryker, Terry Tollefson, and Margot Stern Strom. "Human Rights Education through the Facing History and Ourselves Program." *Journal of Moral Education* 23 (September 1994): 333–47.

Brabham, Edna Greene. "Holocaust Education: Legislation, Practices, and Literature." *The Social Studies* 88 (May/June 1997): 139–42.

Bradley Commission on History in Schools. *Building a History Curriculum: Guidelines for Teaching History in Schools.* Washington, D.C., 1998.

Browning, Christopher. *Ordinary Men: Reserve Police Battalion 101 and the Final Solution in Poland.* New York: Harper Perennial, 1993.

Broznick, Norman M. "A Theological View of the Holocaust: A Traditional Approach for Traditional Jewish Education." *Jewish Education* 42 (Summer 1973): 13–28.

Butz, Arthur. *The Hoax of the Twentieth Century.* Newport Beach, CA: Noontide Press, 1977.

Cavaiani, Marion. "Yet Another Unfair Demand on Schools." *The Record* (Bergen, NJ), September 15, 1998, L08.

Charny, Israel W. "Teaching the Violence of the Holocaust: A Challenge to Educating Potential Future Oppressors and Victims for Nonviolence." *Jewish Education* 38 (March 1968): 15–24.

Chartock, Roselle Kline. *An Evaluative Study of a Unit Based on the Nazi Holocaust: Implications for the Design of Interdisciplinary Curricula.* Unpublished dissertation, University of Massachusetts, 1979.

———. "Teaching About the Holocaust." *Today's Education* 68 (February–March 1979): 34–37.

Clendinnen, Inga. *Reading the Holocaust.* Cambridge: Cambridge University Press, 1999.

Cohen, Steve. "Just the Facts." In *Remembering the Past, Educating for the Present and the Future: Personal and Pedagogical Stories of Holocaust Educators.* Edited by Samuel Totten. Westport, CT: Praeger, 2002: 81–102.

Cole, Steward G. "Intercultural Education." *Religious Education* 36 (January 1941): 135–86.

Cole, Tim. *Selling the Holocaust: From Auschwitz to Schindler: How History is Bought, Packaged, and Sold.* New York: Routledge, 1999.

Crowe, David M. "The Holocaust, Historiography and History." In *Teaching and Studying the Holocaust.* Edited by Samuel Totten and Stephen Fineberg. Boston: Allyn and Bacon, 2001: 24–61.

Danks, Carol. "An Unlikely Journey: A Gentile's Path to Teaching the Holocaust." In *Remembering the Past, Educating for the Present and the Future: Personal and Pedagogical Stories of Holocaust Educators.* Edited by Samuel Totten. Westport, CT: Praeger, 2002: 81–102.

Dawidowicz, Lucy C. *War against the Jews.* New York: Holt, Rinehart, and Winston, 1975.

———. *The Holocaust and the Historians.* Cambridge: Harvard University Press, 1981.

———. "How They Teach the Holocaust." In *What is the Use of Jewish History?* Edited by Neal Kozodoy. New York: Schocken Books, 1992: 65–83.

———. "Could America Have Rescued the Jews?" In *What is the Use of Jewish History.* Edited by Neal Kozodoy. New York: Schocken Books, 1992: 157–78.

Dewey, John. *Experience and Education.* New York: Touchtone, 1938.

———. "Religion and Morality in a Free Society." In *John Dewey: The Later Works, 1925–1953,* vol. 15. Edited by Jo Ann Boyston. Carbondale, IL: Southern Illinois University Press, 1989: 170–83.

Discussion and Resource Guide—Auschwitz: Inside the Nazi State. Stockbridge, MA: Toby Levine Communications Inc, 2004.

Doneson, Judith E. *The Holocaust in American Film.* Philadelphia: Jewish Public Society, 1987.

Donoho, Grace Ellen. *The Arkansas Holocaust Education Committee's Professional Development Conferences: The Impact on Classroom Implementation.* Unpublished dissertation, University of Arkansas, 1999.

Donvito, Concetta. *A Descriptive Study: The Implementation of the 1994 New Jersey Holocaust/Genocide Mandate in New Jersey Public Middle Schools.* Unpublished dissertation, Seton Hall University, 2003.

Elkin, Sol M. "Minorities in Textbooks: The Last Chapter." *Teachers College Press* 66 (March 1965): 502–8.

Ellison, Jeffrey. *From One Generation to the Next: A Case Study of Holocaust Education in Illinois.* Unpublished dissertation, Florida Atlantic University, 2002.

Epstein, Joel. "Holocaust Education in the United States: The Church Related Institution." In *Remembering of the Future, Working Paper and Addenda: Volume II: The Impact of the Holocaust on the Contemporary World.* Edited by Yehuda Bauer, Alice Eckardt, Franklin H. Littell, Elisabeth Maxwell, Robert Maxwell, and David Patterson. New York: Pergamon Press, 1989: 1168–74.

Estess, Ted L. *Elie Wiesel.* New York: Frederick Ungar Publishing Co., 1980.

Evans, Ronald W. *The Social Studies Wars: What Should We Teach the Children?* New York: Teachers College Press, 2004.

Fackenheim, Emil L. "The Holocaust and Philosophy." *The Journal of Philosophy* 82 (October 1985): 505–14.

———. "Concerning Authentic Responses to the Holocaust." In *The Nazi Holocaust: Perspectives on the Holocaust,* vol. 1. Edited by Michael R. Marrus. Westport: Meckler, 1989: 68–87.

Fallace, Thomas. "Did the Social Studies Really Replace History in American Secondary Schools?" *Teachers College Record* <<110 (9)>> (forthcoming, 2008).

Feingold, Henry. *The Politics of Rescue: The Roosevelt Administration and the Holocaust, 1938–1945.* New Brunswick, NJ: Rutgers University Press, 1970.

Feinstein, Sara. "The Shoah and the Jewish School." *Jewish Education* 34 (Spring 1964): 165–68.

Fine, Melinda. "'You Can't Just Say that the Only Ones Who Can Speak Are Those Who Agree With Your Position': Political Discourse in the Classroom." *Harvard Educational Review* 63 (Winter 1993). Accessed May 21, 2007 from www.edreview.org/harvard93/wi93/w93fin.html.

Fine, Melinda. "Collaborative Innovations: Documentation of the Facing History and Ourselves Program at an Essential School." *Teachers College Record* 94 (Summer 1993): 771–89.

———. *Habits of Mind: Struggling over Values in America's Classrooms*. San Francisco: Jossey-Bass Publishers, 1995.

Fitzgerald, Frances. *America Revised: History Schoolbooks in the Twentieth Century*. Boston: Little, Brown and Company, 1979.

Flunzbaum, Hilene. ed. *Americanization of the Holocaust*. Baltimore: The Johns Hopkins University Press, 1999.

Frampton, Wilson. *A Descriptive Study to Ascertain Curriculum Guidelines for Holocaust Education As Reported By the State Departments of Education*, Unpublished dissertation, Temple University, 1989.

Franciosi, Robert. "Introduction." In *Elie Wiesel: Conversations*. Edited by Robert Franciosi. Jackson: University Press of Mississippi, 2002: ix–xv.

Franklin, Barry and Gary McCulloch. "Introduction." In *The Death of the Comprehensive High School: Historical, Contemporary, and Comparative Perspectives*. Edited by Barry Franklin and Gary McCulloch. New York: PalgraveMacmillan, forthcoming.

Freedman, Theodore, "Introduction: Why Teach About the Holocaust." *Social Education* 42 (April 1978): 263.

Friedlander, Henry. *On the Holocaust: A Critique of the Treatment of the Holocaust in History Textbooks Accompanied By an Annotated Bibliography*. New York: Anti- Defamation League of B'nai B'rith, 1972.

———. "Towards a Methodology of Teaching about the Holocaust." *Teachers College Record* 80 (February 1979): 519–42.

Friedlander, Saul. ed. *Probing the Limits of Representation: Nazism and the "Final Solution."* Cambridge, MA: Harvard University Press, 1992.

Garber, Zev and Bruce Zuckerman. "Why Do We Call 'The Holocaust' The Holocaust? An Inquiry into the Psychology of Labels." *Modern Judaism* 9 (1989): 197–211.

Gardner, Howard. *The Unschooled Mind: How Children Think and How Schools Should Teach*. New York: Basic Books, 1991.

———. *Multiple Intelligences: The Theory into Practice: A Reader*. New York: Basic Books, 1993.

———. *The Disciplined Mind: Beyond Facts and Standardized Tests, The K-12 Education that Every Child Deserves*. New York: Penguin Books, 2000.

Gilbert, Martin. *Auschwitz and the Allies*. New York: Holt Rinehart, and Winston, 1981.

Gilligan, Carol. "In a Different Voice: Woman's Conceptions of Self and of Morality." *Harvard Educational Review* 47 (November 1977): 481–517.

Glanz, Jeffrey. "Ten Suggestions for Teaching the Holocaust." *History Teacher* 32 (August 1999): 547–65.

Glynn, Mary, Geoffrey Bock, and Karen Cohen. *American Youth and The Holocaust: A Study of Four Major Holocaust Curricula*. New York: National Jewish Resource Center, 1982.

Goddard, Henry Herbert. *The Kallikak Family: A Study in the Heredity of Feeble-Mindedness.* New York: The Macmillan Company, 1913.
Goldberg, Mark. "The Holocaust and Education: An Interview with Michael Berenbaum." *Phi Delta Kappan* 79 (December 1997): 317–19.
Goldhagen, Daniel Jonah. *Hitler's Willing Executioners: Ordinary Germans and the Holocaust.* New York: Knopf, 1996.
Goodlad, John. *A Place Called School.* New York: McGraw-Hill, 1984.
Gould, Stephen Jay. *The Mismeasure of Man.* New York: Norton, 1996.
Grant, S.G. *History Lessons: Teaching, Learning, and Testing in U.S. High School Classrooms.* Mahwah, NJ: Lawrence Erlbaum Associates, 2003.
———. ed. *Measuring History: Cases of State-Level testing Across the United States.* Charlotte, NC: Information Age Publishing, 2006.
Harlow, Steve, Rhoda Cummings, and Suzanne M. Aberasturi, "Karl Popper and Jean Piaget: A Rationale for Constructivism" *The Educational Forum* 71 (Fall 2006): 41–48.
Haynes, Stephen R. *Holocaust Education and the Church Related College.* Westport, CT: Greenwood Press, 1997.
———. "Holocaust Education at American Colleges and Universities: A Report on the Current Situation." *Holocaust and Genocide Studies* 12 (Fall 1998): 282–307.
Hector, Susan. "Teaching the Holocaust in England." In *Teaching the Holocaust: Educational Dimensions, Principles and Practice.* Edited by Ian Davies. New York: Continuum, 2000: 105–16.
Hertzberg, Hazel Whitman. *Social Studies Reform 1880–1980.* Boulder, Colorado: SSEC Publications, 1981.
Hilberg, Raul. "Developments in the Historiography of the Holocaust." In *Comprehending the Holocaust: Historical and Literary Research.* Edited by Asher Cohen, Joav Gelber, and Charlotte Wardi. New York: Verlag Peter Lang, 1988: 21–44.
Hirsch, E.D. Jr. "Cultural Literacy." *American Scholar* 52 (1983): 159–69.
———. *Cultural Literacy: What Every American Needs to Know.* Boston: Houghton Mifflin, 1987.
Holt, Evelyn. *Implementation of Indiana's Resolution to Holocaust Education by Selected Language Arts and Social Studies Teachers in Middle Schools/Junior High and High Schools.* Unpublished dissertation, Indiana State University, 2001.
Insdorf, Annette. *Indelible Shadows: Film and the Holocaust.* New York: Random House, 1983.
Jick, Leon A."The Holocaust: Its Use and Abuse within the American Public." *Yad Vashem Studies* 14 (1981): 303–18.
Kane, Michael B. *Minorities in Textbooks: A Study of Their Treatment in Social Studies Texts.* Chicago and New York: Anti-Defamation League of B'nai B'rith, 1970.
Kanter, Leona. "Forgetting to Remember: Presenting the Holocaust in American College Social Science and History Textbooks." ERIC-CRESS ED 439 039 (1998): 1–59.

Katz, Steven. *The Holocaust in Historical Context,* vol. 1. New York: Oxford University Press, 1994.

Kegan, Robert. *In Over Our Heads: The Mental Demands of Modern Life.* Cambridge, MA: Harvard, 1994.

Keith, Sherry. "Politics of Textbook Selection." ERIC project report No. 81-A7 (April 1981).

Kliebard, Herbert M. "Constructing a History of the American Curriculum." In *The Handbook of Research on Curriculum.* Edited by Philip W. Jackson. New York: MacMillan, 1992: 157–84.

———. *The Struggle for the American Curriculum 1893–1958.* New York: Routledge, 1995.

Koka, Jergen. "German History before Hitler: The Debate about the German Sonderweg." *Journal of Contemporary History* 23 (January 1988): 3–16.

Kohlberg, Lawrence. "Stage and Sequence: The Cognitive-Developmental Approach to Socialization." In *Handbook of Socialization Theory and Research.* Edited by David Goslin. Chicago: Rand McNally College Publishing Company, 1969: 347–480.

———. "The Cognitive Developmental Approach to Moral Education." In *Moral Education ... It Comes with the Territory.* Edited by David Purpel and Kevin Ryan. Berkeley: McCutchen Publishing Corporation, 1976: 176–95.

Kohlberg, Lawrence and Rochelle Mayer. "Development as the Aim of Education." *Harvard Educational Review* 42 (Fall 1966): 449–96.

Kolbert, Jack. *The Worlds of Elie Wiesel: An Overview of his Career and his Major Themes.* London: Associated University Press, 2002.

Korman, Gerd. "The Holocaust in American Historical Writing." *Societas—A Review of Social History* (Summer 1972): 251–70.

Krathwohl, D.R., B.S. Bloom, and B.B. Masia. *Taxonomy of Educational Objectives, Handbook II: The Affective Domain.* New York: David McKay, 1964.

Krefetz, Gerald. "Nazism: The Textbook Treatment." *Congress Bi-Weekly* 23 (November 13, 1961): 5–7.

Lagermann, Ellen Condliffe. *An Elusive Science: The Troubling History of Education Research.* Chicago: University of Chicago Press, 2000.

Lang, Berel. *Act and Idea in the Nazi Genocide.* Chicago: University of Chicago Press, 1990.

———. *The Future of the Holocaust: Between History and Memory.* Ithaca, NY: Cornell University Press, 1999.

Levstik, Linda and Christine C. Pappas. "New Directions for Studying Historical Understanding." *Theory and Research in Social Education* 20 (Fall 1992): 369–85.

Lieberman, Marcus. "Facing History and Ourselves a Project Evaluation." *Moral Education Forum* 6 (1981): 36–41.

Lindquist, David Hays. *Towards a Pedagogy of the Holocaust Perspectives of Exemplary Teachers.* Unpublished dissertation, Indiana University, 2002.

Linenthal, Edward T. "The Boundaries of Memory: The United States Holocaust Memorial Museum." *American Quarterly* 46 (September 1994): 406–33.

———. *Preserving Memory: The Struggle to Create America's Holocaust Museum.* New York: Viking, 1995.

Lipstadt, Deborah. E. *Beyond Belief: The American Press and the Coming of the Holocaust 1933–1945.* New York: The Free Press, 1986.

———. *Denying the Holocaust: The Growing Assault on Truth and Memory.* New York: Plume 1994.

———. "Not Facing History." *The New Republic* (March 6, 1995): 26–29.

———. "America and the Memory of the Holocaust, 1950–1965." *Modern Judaism* 16 (1996): 195–214.

Littell, Marcia Sachs, "Breaking the Silence: A History of Holocaust Education in America." In *Remembrance, Repentance, Reconciliation: The 25th Anniversary Volume of the Annual Scholars' Conference on the Holocaust and Churches.* Edited by Douglas F. Tobler. New York: University Press of America, 1998: 195–212.

Locke, Hubert G. "The Holocaust and the American University: Observations on the Teaching and Research in a Graduate Professional Field." In *Remembering of the Future, Working Paper and Addenda: Volume II: The Impact of the Holocaust on the Contemporary World.* Edited by Yehuda Bauer, Alice Eckardt, Franklin H. Littell, Elisabeth Maxwell, Robert Maxwell, and David Patterson. New York: Pergamon Press, 1989: 1188–93.

Loewen, James. *Lies My Teacher Told Me: Everything Your American History Textbook Got Wrong.* New York: Touchtone, 1995.

Loshitzky, Yosefa, ed. *Spielberg's Holocaust: Critical Perspectives on Schindler's List.* Bloomington: Indiana University Press, 1997.

Magurshak, Dan. "The 'Incomprehensibility' of the Holocaust: Tightening Up Some Loose Usage." In *The Nazi Holocaust: Perspectives on the Holocaust,* vol. 1. Edited by Michael R. Marrus. Westport: Meckler, 1989: 88–117.

Maier, Charles S. *The Unmasterable Past: History, Historians, and the German National Identity.* Cambridge: Harvard University Press, 1997.

Marcus, Lloyd. *The Treatment of Minorities in Secondary School Textbooks.* New York: Anti-Defamation League of B'nai B'rith, 1961.

Margalit, Avishni and Gabriel Motzkin. "The Uniqueness of the Holocaust." *Philosophy and Public Affairs* 25 (Winter 1996): 65–83.

Marker, Gerald and Howard Mehlinger. "Social Studies." In *The Handbook of Research on Curriculum.* Edited by. Philip W. Jackson. New York: Macmillan, 1992: 830–51.

McClellan, Edward. *Moral Education in America: Schools and the Shaping of Values from Colonial Times to the Present.* New York: Teachers College Press, 1999.

Mintz, Alan. *Popular Culture and the Shaping of Holocaust Memory in America.* Seattle: University of Washington Press, 2001.

Mitchell, Julie Patterson. *Methods of Teaching the Holocaust to Secondary Students As Implemented by Tennessee Recipients of the Betz-Lippman Tennessee Holocaust Educators of the Year Awards.* Unpublished dissertation, East Tennessee University, 2004.

Moreau, Joseph. *Schoolbook Nation: Conflicts over American History Textbooks from Civil War to Present.* Ann Arbor: University of Michigan, 2004.

Morse, Arthur. *While Six Million Dies: A Chronicle of American Apathy.* New York: Random House, 1968.

Nash, Gary, Charlotte Crabtree, and Ross E. Dunn. *History on Trial: Culture Wars and the Teaching of the Past.* New York, Vintage, 1997.

Newmann Fred M. with the assistance of Donald W. Oliver. *Clarifying Public Controversy: An Approach to Teaching the Social Studies.* Boston: Little Brown, 1970. Novick, Peter. "Holocaust Memory in America." In *The Art of Memory: Holocaust Memorials in History.* Edited by James Young. New York: Prestel-Verlag, 1994: 159–65.

———. *The Holocaust in American Life.* Boston: Houghton Mifflin, 1999.

Oliver, Donald W. and James P. Shaver. *Teaching Public Issues in the High School.* Boston: Houghton Mifflin Company, 1966.

Osborne, Ken. "Fred Morrow and the Source-Method of Teaching History." *Theory and Research in Social Education* 31 (Fall 2003): 466–501.

Parsons, William S. and Samuel Totten. *Guidelines for Teaching about the Holocaust.* Washington, D.C.: United States Holocaust Memorial Museum, 1993.

Pate, Glenn A. "The Holocaust in American Textbooks." In *The Treatment of the Holocaust in Textbooks.* Edited by Randolph Braham. New York: Columbia University Press, 1987: 231–333.

Penkower, Monty Noam. *The Jews wWere Expendable: Free World Diplomacy and the Holocaust.* Urbana: University of Illinois Press, 1983.

Pilch, Judah. "The Shoah and the Jewish School." *Jewish Education* 34 (Spring 1964): 162–65.

Polikov, Leon. *Harvest of Hate: The Nazi Program for the Destruction of the Jews of Europe.* New York: Syracuse University Press, 1954.

Powell, Arthur, Eleanor Farrar, and David Cohen. *The Shopping Mall High School.* Boston: Houghton Mifflin, 1985.

President's Commission on the Holocaust. *Report to the President.* Washington, D.C., 1979.

Proctor, Robert N. *Racial Hygiene: Medicine Under the Nazis.* Cambridge: Harvard University Press, 1988.

Rabinsky. Leatrice. "A Journey Through Memory." In *Remembering the Past, Educating for the Present and the Future: Personal and Pedagogical Stories Of Holocaust Educators.* Edited by Samuel Totten. Westport, CT: Praeger, 2002: 123–35.

Rathenow, Hanns-Fred. "Teaching the Holocaust in Germany." In *Teaching the Holocaust: Educational Dimensions, Principles and Practice.* Edited by Ian Davies. New York: Continuum, 2000: 63–76.

Raths, L.E., M. Harmin, and S.B. Simon. *Values and Teaching: Working with Values in the Classroom.* Columbus, OH: Charles E. Merrill, 1966.

Ravitch, Diane. *The Troubled Crusade.* New York: Basic Books: 1983.

———. "Who Prepares Our History Teachers? Who Should Prepare Our History Teachers?" *The History Teacher* 31 (August 1998): 495–503.

———. *Left Back: A Century of School Reform.* New York: Touchtone, 2000.

———. *The Language Police.* New York: Alfred A Knopf, 2003.

Ravitch, Diane and Chester E. Finn, *What Do Our 17-Year Olds Know? A Report on the First National Assessment of History and Literature.* New York: Perennial Library, 1987.

Reitlinger, Gerald. *The Final Solution.* New York: Beechhurst Press, 1953.

Riley, Karen. "The Holocaust and Historical Empathy." In *Historical Empathy and Perspective Taking in the Social Studies.* Edited by O.L. Davis, Elizabeth A. Yeager, and Stuart Foster. Lanham, MD: Rowman and Littlefield, 2000: 239–66.

Riley, Karen and Samuel Totten. "Understanding Matters: Holocaust Curricula and the Social Studies Classroom." *Theory and Research in Social Education* 30 (Fall 2002): 541–62.

Rock, Tracy, Tina Heafner, Katherine O'Connor, Jeff Passe, Sandra Oldendorf, Amy Good, and Sandra Byrd. "One State Closer to a National Crisis: A Report on Elementary Social Studies Education in North Carolina Schools." *Theory and Research in Social Education* 34 (Fall 2006): 455–83.

Roskies, Diane. *Teaching the Holocaust to Children: A Review and Bibliography.* Hoboken, NJ: KTAV Publishing House, INC, 1975.

Schlesinger, Jr., Arthur. *The Disuniting of America: Reflections on a Multicultural Society.* New York: W.W. Norton, 1991.

Schultz, Lynn Hickey, Dennis J. Barr, and Robert L. Selman. "The Value of a Developmental Approach to Evaluating Character Development Programmes: An Outcome Study of Facing History and Ourselves." *Journal of Moral Education* 30 (March 2001): 3–27.

Schweber, Simone. *Teaching History, Teaching Morality: Holocaust Education in American Public Schools.* Unpublished dissertation, Stanford University, 1998.

———. "Simulating Survival." *Curriculum Inquiry* 33 (Spring 2003): 139–88.

———. *Making Sense of the Holocaust: Lessons from Classroom Practice.* New York:Teachers College Press, 2003.

———. "Holocaust Fatigue in Teaching Today." *Social Education* 70 (January 2006): 44–49.

———. "Holocaust Education at Lubavitch Girls Yeshiva." Unpublished manuscript version (forthcoming in *Jewish Social Studies*).

———. "What Happened to Their Pets?: Third Graders Encounter the Holocaust." Unpublished manuscript version (forthcoming *Teachers College Record*).

Schweber, Simone and Rebekah Irwin. "'Especially Special': Leaning About Jews in a Fundamentalist Christian School." *Teachers College Record* 105 (December 2003): 1693–719.

Sepinwall, Harriet. "Incorporating Holocaust Education into K-4 Curriculum and Teaching in the United States." *Social Studies and the Young Learner* (January/February 1999): 5–8.

Shandler, Jeffrey. "Aliens in the Wasteland: America Encounters with the Holocaust on 1960s Science Fiction Television." In *The Americanization of the Holocaust.* Edited by Hilene Flanzbaum. Baltimore: Johns Hopkins University Press, 1999: 33–44.

Shandler, Jeffrey. *While America Watches: Televising the Holocaust.* New York: Oxford University Press, 1999.

Shandley, Robert. ed. *Unwilling Germans?: The Goldhagen Debate.* Minneapolis: University of Minnesota Press, 1998.

Shawn, Karen. "Current Issues in Holocaust Education." *Dimensions* 9 (1995): 15–18.

Sheramy, Rona. *Defining Lessons: The Holocaust in American Jewish Education.* Unpublished dissertation, Brandeis University, 2000.

Short, Geoffrey and Carole Ann Reed. *Issues in Holocaust Education.* Burlington, VT: Ashgate, 2004.

Simon, Sidney B., Leland W. Howe, and Howard Kirschenbaum. *Values Clarification: A Handbook of Practical Strategies for Teachers and Students.* New York: Hart Publishing Company Inc., 1972.

Sizer, Theodore. *Horace's Compromise: The Dilemma of the American High School.* Boston: Houghton Mifflin, 1984.

Smith, Tom W. *Holocaust Denial: What the Survey Data Reveal.* New York:American Jewish Committee, 1995.

Spector, Karen. "God on the Gallows: Reading the Holocaust though Narratives of Redemption." *Research in the Teaching of English* 42 (August 2007): 7–55.

Spiegelman, Marvin J. "On the Holocaust and Jewish Education," *Jewish Education* 43 (Spring 1964): 36–37.

Stern Strom, Margot. "A Work in Progress." In *Working to Make a Difference: The Personal and Pedagogical Stories of Holocaust Educators Across the Globe.* Edited by Samuel Totten. Lanham, MD: Rowman and Littlefield, 2003: 69–102.

Thayer, Louis. ed. *Affective Education: Strategies for Experiential Learning.* La Jolla, CA: University Associates, 1976.

Totten, Samuel. "A Holocaust Curriculum Evaluation Instrument: Admirable Aim, Poor Result." *Journal of Curriculum and Supervision* 13 (Winter 1998): 148–66.

———. "The Start is as Important as the Finish: Establishing a Foundation for Study of the Holocaust." *Social Education* 62 (February 1998): 70–76.

———. "Diminishing the Complexity and Horror of the Holocaust: Using Simulations in an Attempt to Convey Historical Experiences." *Social Education* 64 (April 2000): 165–71.

———. "Holocaust Education in the United States." In *Holocaust Encyclopedia.* Edited by Walter Lacqueur. New Haven: Yale University Press, 2001: 305–12.

———. *Holocaust Education: Issues and Approaches.* Boston: Allyn and Bacon, 2002.

———. "Why?" In *Remembering the Past, Educating for the Present and the Future: Personal and Pedagogical Stories of Holocaust Educators.* Edited by Samuel Totten. Westport, CT: Praeger, 2002: 177–222.

Totten, Samuel and Karen L. Riley. "Authentic Pedagogy and the Holocaust: A Critical Review of State Sponsored Holocaust Curricula." *Theory and Research in Social Education* 33 (Winter 2005): 120–41.

Toubin, Isaac. "How to Teach the Shoah." *Conservative Judaism* 18 (Summer, 1964): 22–26.

Tyack, David. *Seeking Common Ground: Public Schools in a Diverse Society* Cambridge: Harvard, 2004.

Ury, Zalman. "The Shoah and the Jewish School." *Jewish Education* 34 (Spring 1964): 168–72.

VanFossen, Philip. "'Reading and Math Take So Much Time…' An Overview of Social Studies Instruction in Elementary Classrooms in Indiana." *Theory and Research in Social Education* 33 (Summer 2005): 376–403.

VanSledright, Bruce. *In Search of America's Past: Learning to Read History in Elementary School*. New York: Teachers College Press, 2002.

Wegner, Gregory. "What Lessons Are There from the Holocaust for My Generation Today? Perspectives on Civic Virtue from Middle School Youth." *Journal of Curriculum and Supervision* 13 (Winter 1998): 167–83.

Weinstein, G. and M.D. Fantini. *Toward Humanistic Education: A Curriculum of Affect*. New York: Praeger, 1970.

Wiesel, Elie. "Then and Now: The Experiences of a Teacher." *Social Education* 42 (April 1978): 266–76.

———. "Trivializing the Holocaust: Semi-Fact and Semi-Fiction." *New York Times*, April 16, 1978, B1.

———. *All Rivers Run to the Sea: Memoirs*. New York: Alfred A. Knopf, 1995.

———. *And the Sea is Never Full: Memoirs, 1969–* . New York: Alfred Knopf, 1999.

———. *Elie Wiesel: Conversations*. Edited by Robert Franciosi. Jackson: University of Mississippi Press, 2002.

Wineburg, Samuel. "Reading Historical Texts: Notes on the Breach between School and Academy." *American Educational Research Journal* 28 (1991): 495–519.

———. *Historical Thinking and Other Unnatural Acts: Charting the Future of Teaching the Past*. Philadelphia: Temple University, 2001.

Witt, Joyce Arlene. *A Humanities Approach to the Study of the Holocaust: A Curriculum for Grades 7–12*. Unpublished dissertation, Illinois State University, 2000.

Wraga, William G. *Democracy's High School: The Comprehensive High School and Educational Reform in the United States*. Landham, MA: University Press of America, 1994.

Wyman, David S. *The Abandonment of the Jews: America and the Holocaust 1941–1945*. New York: Pantheon Books, 1984.

Yeager, Elizabeth Anne and O.L. Davis. eds. *Wise Social Studies Teaching in an Age of High-Stakes Testing: Essays on Classroom Practices and Possibilities*. Charlotte, NC: Information Age Publishing, 2005.

Young, James E. *The Texture of Memory: Holocaust Memorials and Meaning*. New Haven: Yale University Press, 1993.

———. ed. *The Art of Memory: Holocaust Memorials in History*. New York: Prestel-Verlag, 1994.

Young, James E. "Towards a Received History of the Holocaust." *History and Theory* 36 (December 1997): 36–37.

———. "America's Holocaust: Memory and the Politics of Identity." In *The Americanization of the Holocaust.* Edited by Hilene Flunzbaum. Baltimore: The Johns Hopkins University Press, 1999: 68–82.

Zimmerman, Jonathan. *Whose America? Culture Wars in the Public Schools.* Cambridge: Harvard, 2003.

Index

The Emergence of Holocaust Education in American Schools

SECONDARY EDUCATION IN A CHANGING WORLD

Series editors: Barry M. Franklin and Gary McCulloch

Published by Palgrave Macmillan:

The Comprehensive Public High School: Historical Perspectives
By Geoffrey Sherington and Craig Campbell
(2006)

Cyril Norwood and the Ideal of Secondary Education
By Gary McCulloch
(2007)

The Death of the Comprehensive High School?: Historical, Contemporary, and Comparative Perspectives
Edited by Barry M. Franklin and Gary McCulloch
(2007)

The Emergence of Holocaust Education in American Schools
By Thomas D. Fallace
(2008)